SpringerBriefs in Environmental Science

SpringerBriefs in Environmental Science present concise summaries of cutting-edge research and practical applications across a wide spectrum of environmental fields, with fast turnaround time to publication. Featuring compact volumes of 50 to 125 pages, the series covers a range of content from professional to academic. Monographs of new material are considered for the SpringerBriefs in Environmental Science series.

Typical topics might include: a timely report of state-of-the-art analytical techniques, a bridge between new research results, as published in journal articles and a contextual literature review, a snapshot of a hot or emerging topic, an in-depth case study or technical example, a presentation of core concepts that students must understand in order to make independent contributions, best practices or protocols to be followed, a series of short case studies/debates highlighting a specific angle.

SpringerBriefs in Environmental Science allow authors to present their ideas and readers to absorb them with minimal time investment. Both solicited and unsolicited manuscripts are considered for publication.

More information about this series at http://www.springer.com/series/8868

Beidou Xi · Yonghai Jiang
Mingxiao Li · Yu Yang · Caihong Huang

Optimization of Solid Waste Conversion Process and Risk Control of Groundwater Pollution

Springer

Beidou Xi
Chinese Research Academy
of Environmental Sciences
Beijing
China

Yu Yang
Chinese Research Academy
of Environmental Sciences
Beijing
China

Yonghai Jiang
Chinese Research Academy
of Environmental Sciences
Beijing
China

Caihong Huang
Chinese Research Academy
of Environmental Sciences
Beijing
China

Mingxiao Li
Chinese Research Academy
of Environmental Sciences
Beijing
China

ISSN 2191-5547 ISSN 2191-5555 (electronic)
SpringerBriefs in Environmental Science
ISBN 978-3-662-49460-8 ISBN 978-3-662-49462-2 (eBook)
DOI 10.1007/978-3-662-49462-2

Library of Congress Control Number: 2016932350

Printed on acid-free paper

This Springer imprint is published by SpringerNature
The registered company is Springer-Verlag GmbH Berlin Heidelberg

Contents

Chapter 1
Introduction

Abstract The book proposes an optimal scheme for full-life-cycle solid waste management that addresses the critical technological bottlenecks, and introduces a new technical model that reflects the synergies of resource utilization and secondary pollution control. A set of technical methods has been set out for scientific and reasonable on-site investigation, risk assessment, management and control, to minimize the risk of groundwater contamination by solid waste at landfill site. This book will provide useful guidance on optimal management of solid waste and risk assessment and management of groundwater pollution at landfill site for government managers, environmental researchers, and people involved in and concerned about solid waste disposal.

Keywords Solid waste · Resource recovery · Secondary pollution control · Groundwater · Investigation method · Risk ranking · Management

Solid waste includes garbage, sludge, food wastes, straw, manure, electronic waste, construction waste, and medical waste. It contains large amounts of toxic and hazardous substances and therefore poses a great threat to the human environment. A systemic study of the metabolic process of solid waste, law of conversion of major pollutants, and their environmental effects will facilitate the solution to solid waste pollution, recycling pathways, and strategies for multi-objective control and management. Furthermore, this will reduce the negative impact of solid waste on the environment, improve the ecological environment, and lay an important foundation for developing the venous industry. As a metabolite of lives, solid waste are complex, numerous, and widespread, of which the large amount of recyclable materials are potential resources for utilization, but heavy metals, perishable organic matter (POM), and pathogenic microorganisms, if improperly disposed, will cause serious pollution. Effective utilization of solid waste and control of secondary pollution has become a technical and social problem that must be addressed in urban development and rapid urbanization.

Focusing on solid waste, the book probes into the law of material conversion and builds a metabolic dynamics model integrating different disposal patterns of typical

B. Xi et al., *Optimization of Solid Waste Conversion Process and Risk Control of Groundwater Pollution*, SpringerBriefs in Environmental Science,
DOI 10.1007/978-3-662-49462-2_1

pollutants, in the hope of identifying metabolism mechanisms, migration pathways, and major influencing factors in solid waste treatment and disposal. At the same time, a dynamic multi-objective optimization model reflecting the Chinese characteristics is introduced, which rests on multi-objective planning framework, pollution loss theory, and uncertainty analysis. A mechanism of management toward effective environmental and economic trade-offs is formed, covering optimization, validation, and feedback. On this basis, the book makes an in-depth analysis of solid waste pollution characteristics and resource potential, and expounds on technologies for solid waste resource recovery and secondary pollution control from a systemic and holistic perspective, covering the whole process from waste collection, transportation and mechanical sorting, bioaugmentation and resource recovery, control of secondary pollution, and system integration and management optimization. The idea to maximize reuse and recycling and reduce landfill disposal by classifying and sorting waste: biofortification and humification of organic components, pyrolysis and gasification of combustible components, and recycling of other plastics, metals, glass, and paper.

Landfill, as the final disposal of solid waste, inevitably causes a certain degree of perimeter groundwater pollution in the course of operation, due to such factors as construction, barrier layer damage, and uneven subsidence. Investigation and risk assessment of groundwater contamination is an important technical means to understand landfill pollution, forecast pollution trends, and assess contamination risk. Based on the findings of domestic and international groundwater investigation, the book elaborates the general procedures for groundwater contamination investigation in landfills of different types and construction and operation models, including collection of basic data, deployment of monitoring sites, selection of monitoring items, and data integration. On this basis, the classical model—Multimedia, Multi-pathway, Multi-receptor Risk Assessment (3MRA)—is optimized for application. Meanwhile, in order to achieve scientific classification and effective management of groundwater contamination at landfill sites and ensure special regulation of landfills prone to large groundwater contamination incidents, the book proposes a suitable index system and methodology for groundwater contamination risk classification, as well as management procedures and methods. Time-based risk control measures and management programs are also provided specific to landfills built and under planning and construction. These will offer a scientific basis and technical support to environmental protection departments for the classified management of groundwater contamination risk in landfills toward groundwater environmental security.

Chapter 2
Solid Waste Conversion and Dynamic Multi-objective Optimization

Abstract A systemic study of the metabolic process of solid waste and the law of conversion and environmental effects of major pollutants will facilitate the solution to solid waste pollution, recycling pathways and multi-objective optimization strategies. Focusing on solid waste, this chapter probes into the law of material conversion, builds a metabolic dynamics model for typical pollutants integrating different disposal patterns, and reveals the metabolism mechanisms, migration pathways and major influencing factors in solid waste treatment and disposal. A mechanism of management towards effective environmental and economic trade-offs is formed, covering optimization, validation, and feedback.

Keywords Solid waste · Material transformation · Dynamics model · Metabolism mechanisms

2.1 Overview

Solid waste consists of electronic waste, construction waste and medical waste, as well as organic waste, including garbage, sludge, food wastes, straw and manure. It contains large amounts of toxic and hazardous substances that threaten the human environment. A systemic study of the metabolic process of solid waste and the law of conversion and environmental effects of major pollutants will facilitate the solution to solid waste pollution, recycling pathways and multi-objective optimization strategies. Furthermore, this will mitigate the negative impact of solid waste on the environment, improve the ecological environment, and lay an important foundation for developing the venous industry. Focusing on solid waste, this chapter probes into the law of material conversion, builds a metabolic dynamics model for typical pollutants integrating different disposal patterns, and reveals the metabolism mechanisms, migration pathways and major influencing factors in solid waste treatment and disposal. Furthermore, a dynamic multi-objective optimization model reflecting the Chinese characteristics is introduced, which rests on

B. Xi et al., *Optimization of Solid Waste Conversion Process and Risk Control of Groundwater Pollution*, SpringerBriefs in Environmental Science,
DOI 10.1007/978-3-662-49462-2_2

multi-objective planning framework, pollution loss theory, and uncertainty analysis. A mechanism of management towards effective environmental and economic trade-offs is formed, covering optimization, validation, and feedback.

2.2 Law of Solid Waste Conversion and the Environmental Effects

2.2.1 Humification of Organic Waste

2.2.1.1 Changes in Physical and Chemical Properties and Biological Booster Doses

In the humification process, compost temperature change, within a certain range, is positively correlated with growth and reproduction of microorganisms, including heating, thermophilic and cooling (mature period) phases. Studies have shown that, in general, humification occurs in the late thermophilic phase. In aerobic humification process, organic macromolecules are degraded by microbes into small molecules and finally converted to CO_2 and H_2O. The NH_3 volatilization first increases slowly and then gradually decreases in the compost heating and thermophilic phases before achieving relative stability in the mature period. The study found the oxygen consumption rate and CO_2 release rate reach the peak in the thermophilic phase. Microbial inoculation can suppress the NH_3 volatilization and accelerate the humification process by significantly increasing the number of microbes and efficiency of organic matter decomposition, as well as oxygen consumption rate and CO_2 release rate in the thermophilic phase.

2.2.1.2 Dynamic Changes of Organic Matter

(1) Total organic carbon (TOC)

Humification is generally accompanied by organic matter mineralization, and the two are a unity of opposites. In this process, a part of organic carbon is mineralized into CO_2 and H_2O, releasing energy to support microbial growth. Organic carbon content shows a decreasing trend, with notable decline in the heating, thermophilic phases and moderate decline in the mature period. Hence, humification is generally considered dominant in the middle and late phases. The decomposition rate of organic carbon is fast because microorganisms prefer easily decomposable, simple organic compounds (soluble sugars, organic acids, starch, and etc.). In the mature period of compost, however, only hardly decomposable organic matter (cellulose, hemicelluloses, lignin, and etc.) are left as carbon source, slowing down the decomposition rate.

(2) Decomposable organic matter (DOM)

In the humification process, decomposable organic matter experiences down—up—down fluctuations, and ends up with a sharp decline when the compost tends to be stable. In the heating phase, the easily decomposable organic matter reduces quickly, given the favorable high oxygen content. In the thermophilic phase, exogenous microbial activities are very active, and with organic matter decomposition, a lot of easily decomposable organic matter is produced, resulting in a rise of net organic matter. To the cooling phase, the microbial decomposition of organic matter weakens and mainly serves the needs of microorganisms themselves, so the content of organic matter tends to decrease. In the mature period, the content of decomposable organic matter stabilizes at a relatively low level.

(3) Dissolved organic carbon (DOC) content

Generally, the DOC content shows a downward trend in the humification process. In the early phase, the DOC concentration is relatively stable because the rapid decomposition of fat and carbohydrate replenishes DOC necessary to microbial activities. As microorganisms multiply rapidly in the composting, entailing large DOC consumption, the DOC concentration is significantly reduced.

(4) DOC composition

The DOC structure in different phase is as shown in Fig. 2.1. In the initial phase of humification, DOC is mainly comprised of simple structural protein, but with the process of decomposition, humic substances increase, making the DOC composition complex (He et al. 2014).

2.2.1.3 Change in Small Molecular Organic Acids (He et al. 2011a)

Organic acids are an important intermediate product in the humification process. A part of them are mineralized in the tricarboxylic acid cycle, providing nutrients for microbial growth, and the other part, retained as humus. It should be noted that volatile fatty acid (VFA) is the main component of malodorous gases. Small molecular organic acids affect pH in the humification process and the accumulation will acidify compost and hamper microbial growth.

(1) Volatile organic acids

Volatile organic acids increase considerably in the initial humification of organic waste, followed by a drastic decline afterwards. The whole humification process sees two peaks of volatile organic acids.

(2) Nonvolatile organic acids

Similar to volatile organic acids, nonvolatile organic acids also peak twice in the composting, though at different times.

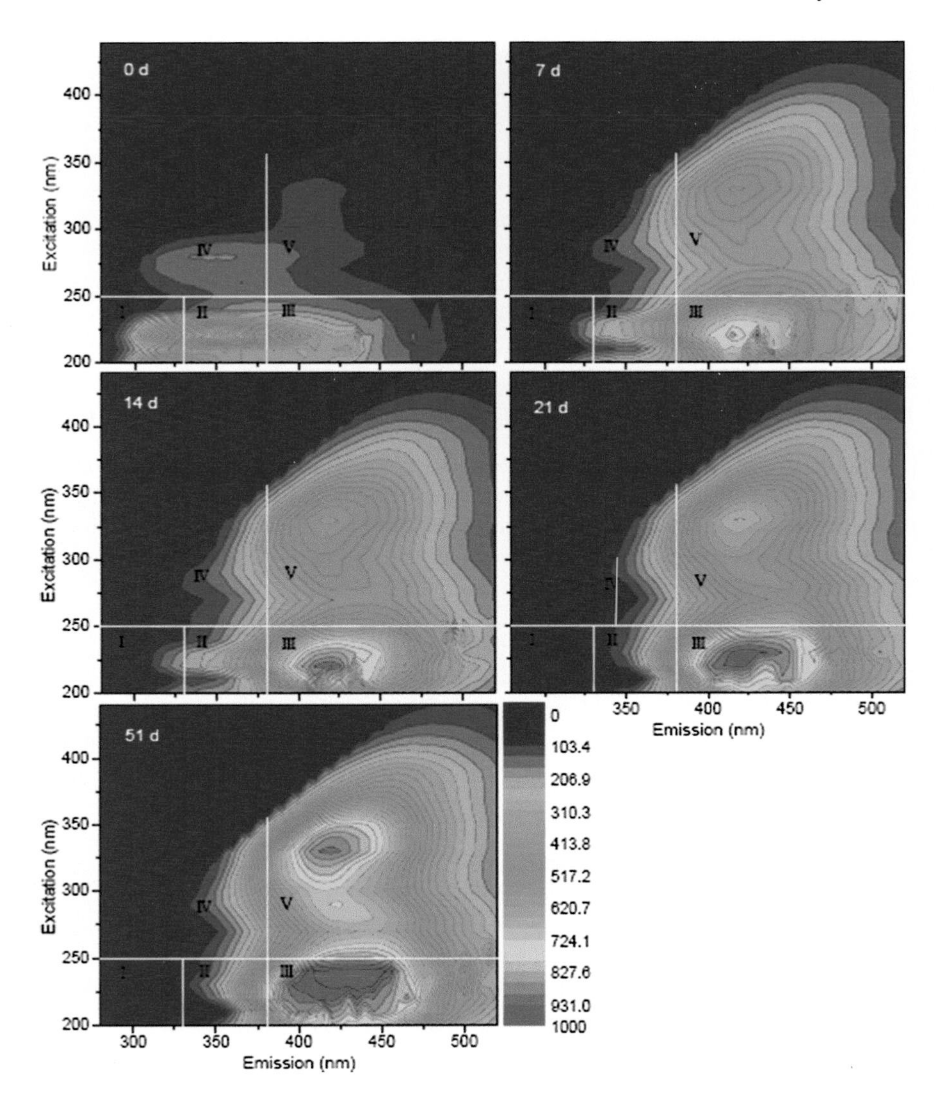

Fig. 2.1 Three-dimensional (3-D) fluorescence spectrograms of the DOC composition "Reprinted from He et al. (2011a, b), with permission from Elsevier"

(3) Total organic acids

In the humification process, bioaugmentation changes greatly the total organic acids. The total organic acids produced in inoculated microbial treatment are much higher than those of exogenous microbial treatment. In late composting, the total amount of organic acids decreases drastically until the mature period, while non-inoculated exogenous microbial treatment maintains a downward trend. Therefore, judging from organic acids, bioaugmentation can facilitate humification.

2.2.1.4 Change in Humus Composition and Characteristics

(1) Humus content

Total humus shows a positive correlation with organic carbon in the humification of organic waste process. In the early and middle phases, the humus content drops fast and exogenous microbial treatment weakens notably more than inoculated microbial treatment. Humic and fulvic acids are important components of humus that play a decisive role in the quality of humus. In the humification process, humic acids first decrease and then increase, while fulvic acids tend to decline. Fulvic acids have relatively low molecular weight and simple molecular structure. In the composting process, partly they are decomposed by microorganisms and partly converted to humic acids of large molecular weight and complex molecular structure.

(2) Degree of humification

Under normal circumstances, the degree of humification of the garbage compost can be represented by three parameters: humification ratio [HR = (HA + FA) × 100/TC, where TC stands for total organic carbon], humification index (HI = HA/FA), and percentage of humic acid (HP = HA × 100/HS). The HR value presents two valleys in the humification process, with an apparent increase in the mature period. Inoculating microbial agents is conducive to the formation of humus and humic acids and further humification, and makes the quality of humic acid superior.

(3) Humic acid spectroscopy

Aromatized humic acids are significantly enhanced with humification. In comparison, exogenous microbial treatment considerably strengthens the condensation of humic acids. As humification stabilizes, humic acids of different sources and treatments have substantially similar infrared spectral shape, but sharply differ in the intensity of absorption in characteristic peaks. This implies a massive impact of different exogenous microbial treatments on structural units and functional group content of humic acids.

2.2.1.5 Dynamic Changes of Organic Nitrogen

The conversion of nitrogen is subject to microbial activities and decides the final maturity. In the humification process, nitrogen is either fixed or released, depending on raw materials (Yang et al. 2006). Nitrogen is one of the main elements of organic waste. Under aerobic humification circumstances, nitrogen exists in the forms of ammonia, nitrate and organic nitrogen, and suffers large losses when the C/N ratio is low. An analysis of different nitrogen-contained matters that reveals material flow and conversion of nitrogen will provide theoretical support for the research to optimize the nitrogen cycle and the retention of nitrogen.

(1) Total nitrogen

In the humification process, the total nitrogen shows a downward trend due to the loss of nitrogen in the decomposition of organic matter. Microbial fermentation involving bioaugmentation only accelerates the composting process and does not cause serious nitrogen loss.

(2) Dissolved organic nitrogen (DON) and insoluble nitrogen

DON accounts for about 75–95 % of and closely correlates with the total nitrogen (He et al. 2015). In the early phase, DON exhibits a downward trend and the treatment of DON with exogenous microorganisms is low, suggesting that exogenous microbes accelerate organic nitrogen mineralization in the humification process.

(3) Amino nitrogen

Amino is a major form of DON. Studies show that amino nitrogen accounts on average for about 33.04 % of and closely correlates with DON. Microbes are conducive to the accumulation of amino nitrogen in the late humification process.

(4) Amide nitrogen

Amide accounts on average for about 18.16 % of DON. In the heating and thermophilic phases, amide nitrogen is increasing, but in the cooling phase, declines significantly until enters a stable state. The entire process sees a notable decrease of amide treatment with exogenous microorganisms, indicating that exogenous microorganisms can reduce the generation of amide nitrogen.

(5) Amino sugar nitrogen

Amino sugar nitrogen, an important component of microbial life, is closely related with microbial biomass. It increases gradually with microbial biomass in the humification process, and drops markedly with the death and decomposition of microorganisms. The content of amino sugar nitrogen tends to stabilize when the compost matures.

(6) Nitrogen in unknown forms

Nitrogen also exists in nucleic acids and their derivatives, phospholipids, vitamins and other derivatives. As composting proceeds, nitrogen in unknown forms decreases, notably in the heating phase. Within 7–63 days, there are complex changes of microbial treatments, but comparatively speaking, exogenous microbial treatment does not cause excessive loss of nitrogen. Exogenous microbial treatment only accelerates the decomposition of organic matter and in a sense, shortens the humification period. Regardless of phase, exogenous microbial treatment of amide nitrogen is always less than non-vaccinated, while amino sugar nitrogen exhibits the opposite trend with amide nitrogen, with no significant differences between the various exogenous microbes.

2.2.1.6 Phosphorus Conversion

(1) Organic phosphorus

Humification is accompanied by organic phosphorus mineralization, but the total amount of organic phosphorus still increases gradually in the humification process. It can be attributed to two reasons: (a) decomposition at a greater rate than mineralization maintains the relative content of organic phosphorus and (b) dissolved phosphorus released in the mineralization has been reused as a component of microbial life in the compost.

(2) Dissolved phosphorus

Dissolved phosphorus first decreases and then increases in the humification process. The content is reduced in the heating phase due to the use of phosphorus for microbial reproduction and metabolism. In the cooling phase, the demand for phosphorus decreases with the end of metabolic process and dead microbes are mineralized, pushing up dissolved phosphorus content in the compost.

(3) Rapidly available phosphorus

Rapidly available phosphorus content shows an upward trend in the humification process. The introduction of insoluble ground rock phosphate to the humification of organic waste will significantly increase the content of rapidly available phosphorus, while low molecular weight organic acids and microbes will accelerate the conversion of ground rock phosphate.

2.2.1.7 Microbial Communities

In the aerobic humification process, cultivable bacteria increase followed by a decrease. In the heating phase when the temperature is relatively low, mesophilic microorganisms, including bacteria and fungi, mainly live in the compost and degrade organic matter. Dow to slow propagation velocity, the fungi population is much smaller than the bacteria population. Generally, microbial communities are related with the organic waste composition and microbial species are very rich. After entering the thermophilic phase when the temperatureis not less than 50 °C, a large number of mesophilic microorganisms die and most fungi turn to spores or die at a temperature greater than 60 °C. Thermophilic bacteria take a dominate position, and thermophilic Bacillus or heat-resistant microorganisms become the primary bacterial population of the compost. In the thermophilic phase, organic matter decomposition speeds up with fast growth, reproduction, and metabolism of bacteria and organic matter soluble in water. In the cooling phase, thermophilic microorganisms gradually lose dominance because of the lack of organic matter. When the compost temperature continues to fall below 50 °C, mesophilic microorganisms become alive again and mesophilic fungi and actinomycetes that degrade macromolecule organic matter become dominant populations (Li et al. 2012).

2.2.2 *Anaerobic Digestion of Organic Waste*

Anaerobic digestion of solid waste is also known as anaerobic fermentation, a process of organic matter decomposition with CO_2 and CH_4 generation under anaerobic conditions by a variety of (anaerobic or facultative anaerobic) microbes (Jia et al. 2014). By means of anaerobic digestion, solid waste can be made harmless, reduced, stabilized and used. Anaerobic digestion is considered a sustainable treatment technology, owning to such advantages as low cost, high efficiency, small footprint, and low energy consumption.

2.2.2.1 Theory

Anaerobic fermentation is generally considered to involve three stages, namely hydrolysis, acidogenesis and methanogenesis (Li et al. 2014). Throughout the whole process, the three bacteria communities (fermentation bacteria, acidogenic and acetogenic bacteria, methanogens) interact with each other, eventually degrading complex organic compounds into CH_4 and H_2.

(1) Hydrolysis

Hydrolysis is the first important process of anaerobic fermentation, in which insoluble complex polymers are broken down to soluble monomers or dimers. Organic polymers with large molecular weight cannot pass through the cell membrane and can be directly used by bacteria only when hydrolyzed by extracellular enzymes into small molecules. The rate and extent of hydrolysis is affected by many factors, including temperature, pH, organic matter components (such as lignin, protein and fat content, and carbohydrates), particle size of organic matter, content of ammonia, and hydraulic retention time. A key pre-condition for hydrolysis is the direct contact between extracellular enzymes and the substrate. Where plant residues serve as the substrate, the extent of cellulose and hemicelluloses wrapped by lignin determines biodegradability.

As far as kitchen waste is concerned, hydrolysis generally performs fast due to high water content, rich organic matter, large proportion of carbohydrates, and low lignin content. Under anaerobic conditions, the protein, which accounts for about 20 % of kitchen waste, is hydrolyzed into polypeptide and amino acid and further acidified to VFA and H_2. Under normal circumstances, the proteolysis rate is very slow because the original protein folder (such as hydrogen bonding) is not sensitive to degradation. In this sense, proteolys is a rate-limiting step for the hydrolysis of kitchen waste with high protein content.

(2) Acidogenesis and acetogenesis

Fermentation refers to the biodegradation process following hydrolysis in which organic molecules can be used as an electron acceptor and at the same time, an electron donor. Acidification is a process that dissolved organic matter (DOM) is converted into VFA-dominated end products.

Acidification end products are determined by anaerobic fermentation conditions, substrate, and microbial populations involved in acidification. They are different, depending on the structure and nature of substrate. The acidification process sees the generation of acetic acid, accompanied by such VFAs as butyric acid and lactic acid which are further metabolized by acetic acid bacteria to acetic acid, H_2 and CO_2. According to end products, fermentation can be divided into three categories: butyric acid fermentation, propionic acid fermentation and ethanol fermentation. The end products of butyric acid fermentation mainly include butyric acid, acetic acid, H_2, CO_2 and a small amount of propionic acid. Propionic acid fermentation ends up with propionic acid and acetic acid, while ethanol fermentation ends up with ethanol, acetic acid, CO_2 and H_2.

(3) Methanogenesis

In the anaerobic digestion, the carboxyl of acetic acid is separated from molecule and converted to CO_2, while methyl is converted to CH_4. The matrix used by methanogens is largely simple carbon compounds containing one or two bonded carbon atoms. In an anaerobic reactor, methanogenic bacteria and methanogenic coccus are dominant species, such as methanosarcina barkeri and methanobacterium sohngenii. When the acetic acid concentration is low, methanobacterium sohngenii grows faster, but given high CH_4 concentration, the bacteria growth increases with the acetic acid concentration. Methanosarcina barkeri is likely to take a dominant position when acetic acid accumulates. Given CO_2 and H_2, CH_4 can also be synthesized by hydrogen-oxidizing methanogens, and when the reactor runs stably, CH_4 generated this way can account for 30 % of the total.

2.2.2.2 Hydrothermal Hydrolysis

(1) Crude protein

Crude protein is one of the target products of biomass value maximization using hydrothermal hydrolysis. LamoolPhak held that in the hydrothermal system, raised processing temperature and shortened processing time is an important condition for obtaining high yields of protein. Watchararuji found that protein solubility in water enhances with ionization of water, while the degree of ionization increases with temperature. Compared with the control group, the crude protein content increases at 80 and 120 °C, but decreases at 150 and 200 °C. It means that such a temperature as 80 and 120 °C makes crude protein colloids hydrolyzed, but does not affect the high-level structure of protein, resulting in a lower degree of hydrolysis. Under the above described conditions, such macromolecules as carbohydrates and lipids are liquefied or leached more than crude protein, pushing up the crude protein content. As the temperature continues to rise, changing the internal structure and affecting the peptide bond, the protein solubility is significantly enhanced, so the crude protein content is relatively reduced at 150 and 200 °C.

(2) VFAs

The total amount of liquid VFA increases markedly after hydrothermal hydrolysis. It indicates that with the enhancement of water ionization, the degree of hydrolysis of macromolecular protein substances increases, lowering down the content of small molecular organic acids in the degradation products. Studies found that small molecules formed in a hydrothermal system have a certain catalytic effect on degradation of macromolecules. The VFA content is more impacted by temperature than time and water. In terms of the VFA composition, hydrothermal hydrolysis greatly affects acetic acid, butyric acid and ethanol. Compared with the control group, acetic acid and butyric acid are noticeably higher, while the ethanol content decline substantially. It implies that hydrothermal hydrolysis tends to convert organic matter to acetic acid.

(3) Crude fat and oil slick

Oil slick is an important parameter affecting garbage biochemical pathways. It contains a large amount of long-chain fatty acids (LCFAs) which are adsorbed on the cell surface to limit the transport of nutrients to cells, thus inhibiting microbial growth and impairing the subsequent biochemical treatment efficiency. As far as kitchen waste is concerned, oil slick is closely related to the crude fat content and can be dramatically increased after hydrothermal hydrolysis. Crude fat can be hydrolyzed to glycerin and fatty acids which further facilitate hydrolysis by esterification with sugar. Meanwhile, hydrothermal hydrolysis enhances the diffusion of solid fat of kitchen waste in water and accelerates the formation of oil slick, lowering down the solid fat content.

(4) Sugar

Sugar is an important carbon source for microbial growth. It can also be directly used or converted into other materials or energy. It is mainly derived from carbohydrate hydrolysis, in which carbohydrate collides with hydronium ions (H_3O^+) or hydroxide (OH^-) and breaks down into monomers, i.e. reducing sugar, when the glycosidic bond is broken. In a hydrothermal system, the average kinetic energy of motion of molecule increases with temperature, and intensifies the ionization of water, exerting a larger effect on the glycosidic bond. When the temperature reaches 150 and 200 °C, carbohydrates are highly liquefied, increasing the content of reducing sugar. The total sugar content has the consistent trend with that of reducing sugar.

2.2.2.3 Intermediates (VFAs and Ethanol)

The different processes in anaerobic digestion exert a significant impact on the composition and conversion of VFAs in fermentation.

VFAs are an important intermediate product in the anaerobic metabolic process. Studies found that, the concentration of acetic acid first increases and then decreases, and peaks in the second day in the groups with heat-moisture treatment, while in the control group, the peak of acetic acid concentration arrives in the third day. Under the conditions of 150 °C, 60 min, and 40 % water, the cumulative amount of acetic acid registers the highest of 11,243.66 mg/L and the cumulative H_2 amount also reaches the peak. This further proves that heat-moisture treatment accelerates the rate of hydrolysis and acidification and improves the efficiency of H_2 production. In addition, under these conditions, the propionic acid level is higher than that of other experimental groups. Among VFAs, propionic acid, with the slowest conversion to acetic acid and supreme toxicity, seriously restrains the activity of methanogenes and represses the conversion of organic matter to small molecule acids, thus hindering the CH_4 generation. Butyric acid exhibits similar changes with acetic acid. The maximum content, recording 6268.87 mg/L, is also seen under the conditions of 150 °C, 60 min, and 40 % water. Butyric acid fermentation is common in anaerobic fermentation, and it is butyric and acetic acid fermentation that generate H_2. This is positively correlated with the change in H_2 production. In the control group, liquid propionic and butyric acids are least and cumulative acetic acid relatively low, which coupled with highly active methanogenes, gives rise to the highest cumulative CH_4 production.

2.2.2.4 Reducing Sugar

Reducing sugar is sugar that can form aldehyde and ketone groups in an alkaline solution and be oxidized by appropriate agents to aldonic acid, saccharic acid, monosaccharides including glucose, fructose and glyceraldehyde, disaccharides including lactose and maltose, as well as oligosaccharides. Reducing sugar can be directly used by microorganisms and the measurement can facilitate the observation of microbial growth. Hence, the routine measurement of reducing sugar is commonly applied to monitor the microbial growth on large-scale industrial production. In the anaerobic digestion of food waste, the concentration of reducing sugar decreases after the initial increase and reaches the highest in about 150–250 h.

2.2.2.5 Fluorescent Substances

A parallel factor analysis of the 3-D fluorescence spectra of effluent from the anaerobic digestion reactor of food waste is conducted, the results are as shown in Fig. 2.2. For Component 1, excitation occurs at a wavelength of 225 and 280 nm and emission 340 nm, respectively corresponding to tryptophan substances and

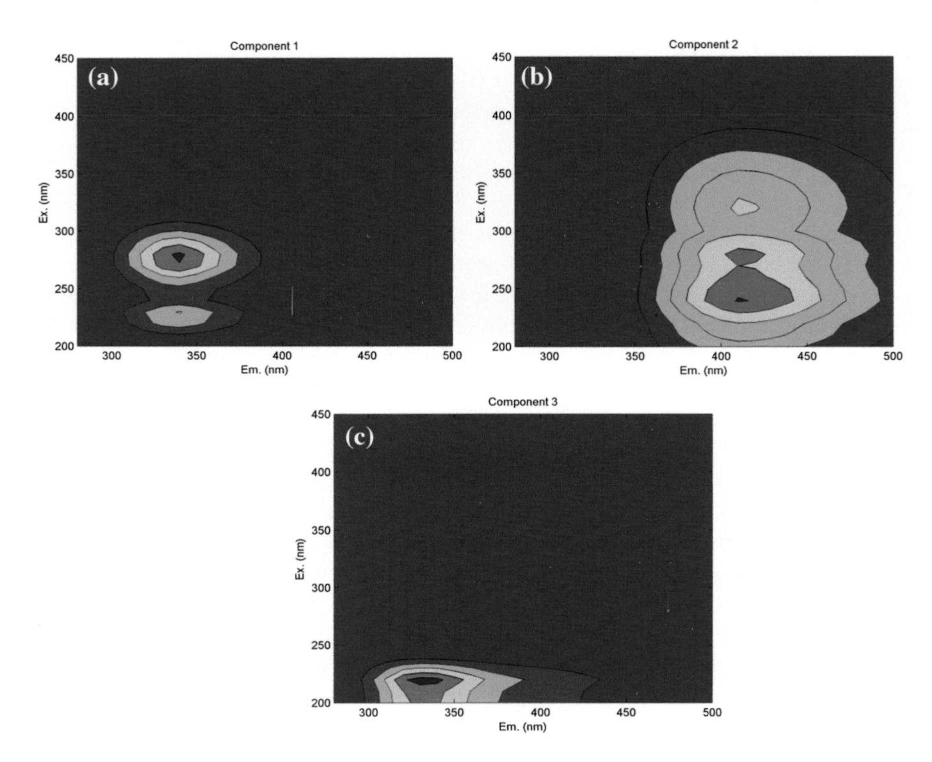

Fig. 2.2 Three-dimensional fluorescence spectrograms by anaerobic fermentation processes

soluble microbial metabolites. For Component 2, excitation occurs at a wavelength of 250 and 330 nm emission 420 nm, which corresponds to the fluorescence contribution of coenzyme NADH and fulvic acids. For Component 3, excitation is positioned at 220 nm and emission 330 nm, corresponding to the fluorescence contribution of protein components.

Judging from the score by fluorescence intensity, Component 2 shows basically the same trend of increase with the duration of anaerobic fermentation in different processes. In anaerobic fermentation process, NADH accumulates in the absence of electron transport chain. Coenzyme NADH has 460 nm UV light fluorescence emission at 340 nm, while the fluorescence for the oxidation state NAD^+ is not observed. In the process of oxidative phosphorylation, NADH transfers electrons to oxygen and is oxidized to NAD^+. The process is inhibited by limited oxygen, leading to NADH accumulation, as fluorescence shown in parallel factor analysis. Accordingly, such characteristics can be applied and monitored. When the reactor is not confined strictly and the anaerobic environment is destroyed, the NADH fluorescence will suddenly reduce.

2.2.3 Solid Waste Landfill

Landfill is a major way to dispose solid waste, in which the organic components are stabilized under the action of microorganisms. It is very important to study the generation and conversion of landfill gas and leachate which is the core of landfill operation and secondary pollution control (Cai 2003).

2.2.3.1 Landfill Gas Generation, Migration and Transformation

Landfill gas generation is a very complex process. The key lies in the biochemical reaction of anaerobic fermentation and decomposition of organic matter in the waste that produces CH_4 and CO_2. According to the type of reaction, the process of organic matter decomposition can be divided into four stages: aerobic decomposition, anaerobic hydrolysis and acidification, anaerobic aerogenesis, and oxidation.

Landfill gas generation rate is described by the following formula:

$$\alpha_k(t) = \sum_{i=1}^{3} C_{Ti} A_i \lambda_i e^{-\lambda_i t} \tag{2.2.1}$$

where in i represents the three components of waste (easily, moderately, and hardly decomposable substances); α_k indicates the annual gas generation rate for gas k (kg/m^3); C_{Ti} indicates the potential total gas generation rate for gas k in component i (kg/m^3); A_i means the content of component i and λ_i the corresponding gas generation constant (yr-1); t stands for time (yr).

In case of ignoring the diffusion change over time, landfill gas changes can be calculated, using the following formula:

$$\begin{aligned} &\frac{\partial}{\partial X}\left(D_{ekm}\frac{\partial \rho_k}{\partial X}\right) + \frac{\partial}{\partial Y}\left(D_{ekm}\frac{\partial \rho_k}{\partial Y}\right) + \frac{\partial}{\partial Z}\left(D_{ekm}\frac{\partial \rho_k}{\partial Z}\right) + \alpha_k(Z) \\ &= \frac{\partial(V_X\rho_k)}{\partial X} + \frac{\partial(V_Y\rho_k)}{\partial Y} + \frac{\partial(V_Z\rho_k)}{\partial Z} \end{aligned} \tag{2.2.2}$$

where in: $V_{x,y,z}$ represents the velocity in direction x, y, and z respectively; ρ_k stands for the mass concentration of gas k in the gas mixture (kg/m^3).

Based on the basic fluid dynamics in porous media theory, a three-dimensional mathematical model is constructed for landfill gas migration and transformation and the finite element and iterative methods introduced to facilitate calculation. Further, the MATLAB visual simulation is carried out, providing a theoretical basis the utilization of landfill gas and control of secondary pollution. In the simulation of landfill gas generation, migration and transformation in an 8 year old landfill, the total pressure reaches the maximum of 23 Kpa in the middle of the landfill, and decreases gradually from the middle to the edges, indicating that the dominant direction of gas transport is from the boundary toward the center in the case of impermeable boundaries. The distribution of pressure may be due to high gas generation rate with fast microbial degradation of organic waste at high temperature in the middle and low gas production rate at a low ambient temperature.

Landfill gas is composed of CH_4 and CO_2 and a small amount of O_2 and N_2. CH_4 and CO_2 are produced from strong sources, i.e. microbial degradation of waste, and take a large proportion in the total pressure (Fig. 2.3), and N_2 is generated only from a small source and gradually diffused beyond the landfill system. As landfill turns old, gas production first increases and then decreases.

2.2.3.2 Leachate Generation, Migration and Transformation

Landfill leachate is organic wastewater with high organic content, and leachate characteristics vary, depending on the landfill age and leachate treatment stages (Cao et al. 2004; Li et al. 2008). A clear understanding of the characteristics of organic pollutants is important for preventing and controlling the spread of contamination and but also lays the basis for the actual leachate treatment and operating parameters (Chen et al. 2005).

In this study, three-dimensional fluorescence spectroscopy, infrared spectroscopy and elemental analysis are combined to follow up organic matter changes at landfills of different ages and in leachate treatment stages (Fig. 2.4).

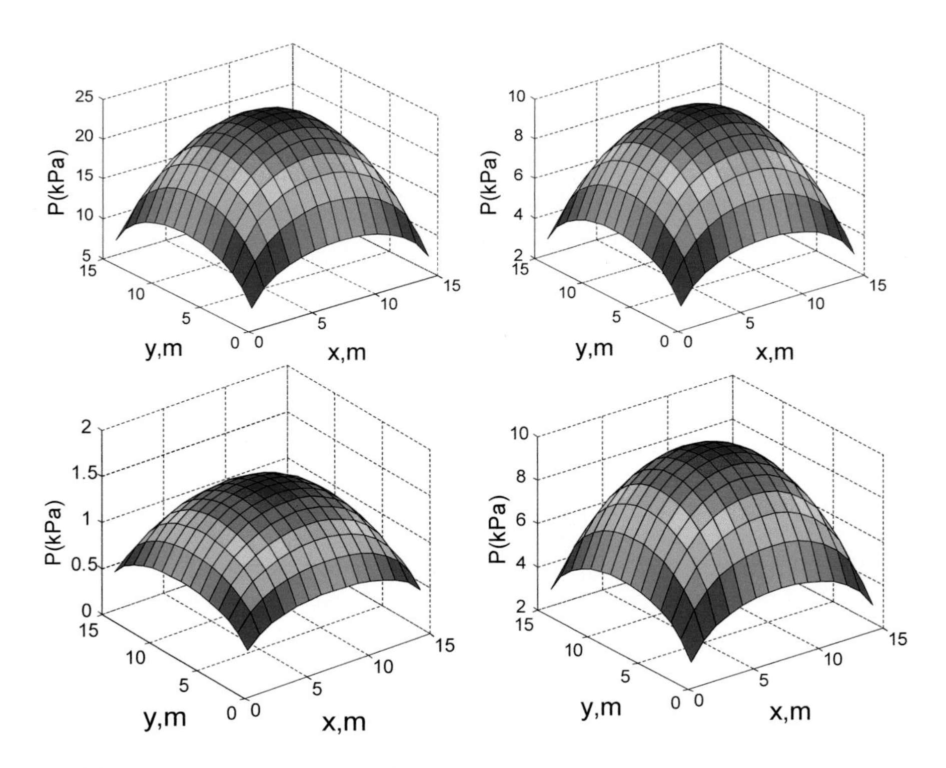

Fig. 2.3 Total pressure changes with depth

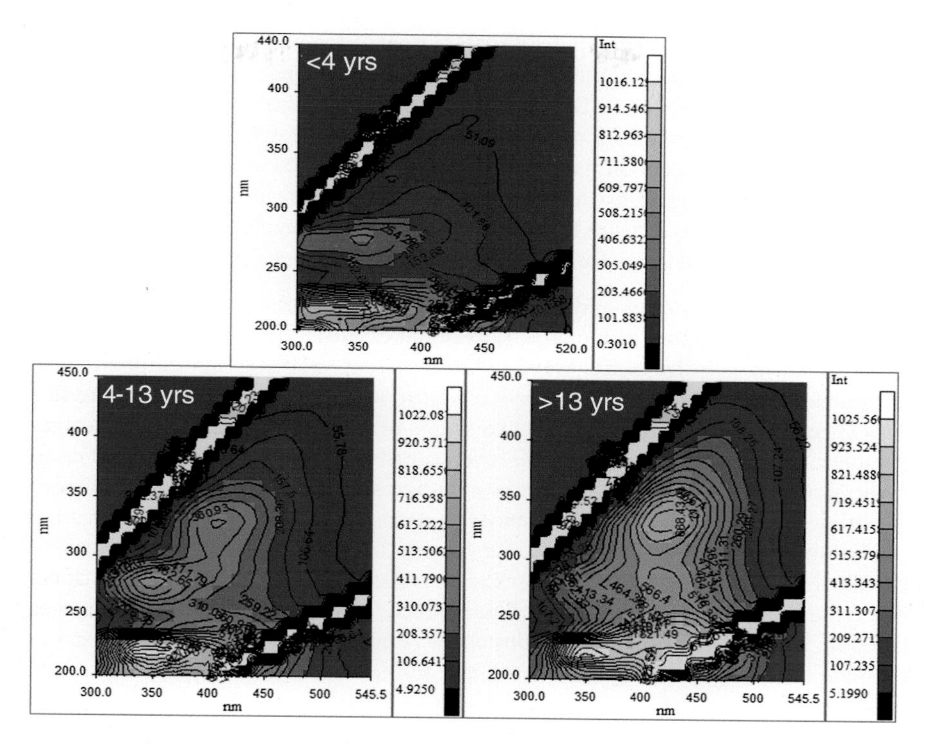

Fig. 2.4 Fluorescence spectrograms for DOM in leachate at landfills of different ages

DOM fluorescence spectra show four kinds of fluorescent peaks including protein-like fluorescence and humic-like fluorescence. At young landfills, the peak occurs in protein-like fluorescence, and DOM mainly consists of simple-structured protein-like substances. At middle-aged and old-age landfills, leachate mainly contains fulvic-like substances and humic-like substances, largely hardly decomposable organic matter. In the landfill reactor simulation, the 3-D fluorescence peak of leachate gradually shifts from protein-like fluorescence to humic-like fluorescence, indicating an increased degree of decomposition.

Protein-like substances can be easily removed through biological processes and humic-like substances through reverse osmosis process. The analysis of fluorescence properties can facilitate the rapid determination of characteristics of organic matter dissolved the leachate sample. Infrared spectroscopy quantifies the changes in properties of organic matter by analyzing chemical groups and bands of substances in the sample and DOM structural changes. The results show that with the humification degree and aromatic series of leachate increase with the age of landfill.

DOM is the most important contaminant of leachate. It can interact with other toxic substances in environmental media, producing composite pollution and impacting migration and transformation of the latter and bioavailability.

(1) DOM in initial landfill leachate mainly contains protein-like substances, but with the extension of landfill life, humic-like substances appear and take an increasing percentage in DOM, and the condensation of molecules increases, strengthening humification. Studies revealed that the specific way of landfill has a notable impact on the DOM composition. In the case of layered landfill, fresh leachate migrates downward along the section, resulting in increased protein-like substances in DOM of underlying aged leachate. This will impact the migration and transformation of pollutants in the media.

In the four DOM groups of different polarities and charges, the hydrophobic acid (HOA) in leachate initially includes tyrosine-like substances, and late, fulvic-like and humic-like substances. Hydrophilic matter (HIM) is embodied in tryptophan-like substances, tyrosine-like substances, fulvic-like and humic-like substances in the early, middle and late stages of landfills respectively, indicating complex structure of organic matter over time. Hydrophobic bases (HOB) and hydrophobic neutrals (HON) remain protein-like substances all the time, but there are fulvic-like and humic-like substances in aged leachate (Fig. 2.5).

Biological treatment is suitable for young landfill leachate, while physico-chemical methods, such as reverse osmosis process, are applicable to aged leachate. The volume ratio of characteristic fluorescences between humic-like substances (Region V + VI) and protein-like substances (Region I + II + III + IV) is used to evaluate the stability of organic matter, and further forecast and optimize leachate treatment techniques (Fig. 2.6).

(2) HOA is the major component of DOM in the leachate, and increases with the age of landfill. Humic-like substances show increased formation capability and weaken the mobility and bioavailability of such environmental pollutants as mercury. HIM decreases over time, of which humic-like substances become less able in the complexation of mercury and reduce the mobility and bioavailability of other toxic substances (Table 2.1). Hence, with the extension of landfill life, the secondary environmental effect of DOM in the leachate abates. HON and HOB of low content have limited impact on DOM's environmental effects.

DOM consists of two typical components: protein-like substances and humic-like substances. Protein-like substances significantly improve the bioavailability of heavy metals and the secondary environmental effects because they are more able in the complexation of heavy metals (mercury) and easy to use and degrade.

(3) The process of bonding DOM and heavy metals in leachate is subject to the pH of media, notably in the acidic and basic solutions. When the pH is lower than 5, the carboxyl group has the dominant role, and when the pH is high ($pH > 9$), the phenolic hydroxyl group plays an important role. When the pH ranges from 6 to 9, β-dicarboxy compounds, alcohols, and inorganic substrate on the surface undego weak dissociation, so there will be little change in the organic matter's ability to bond heavy metals.

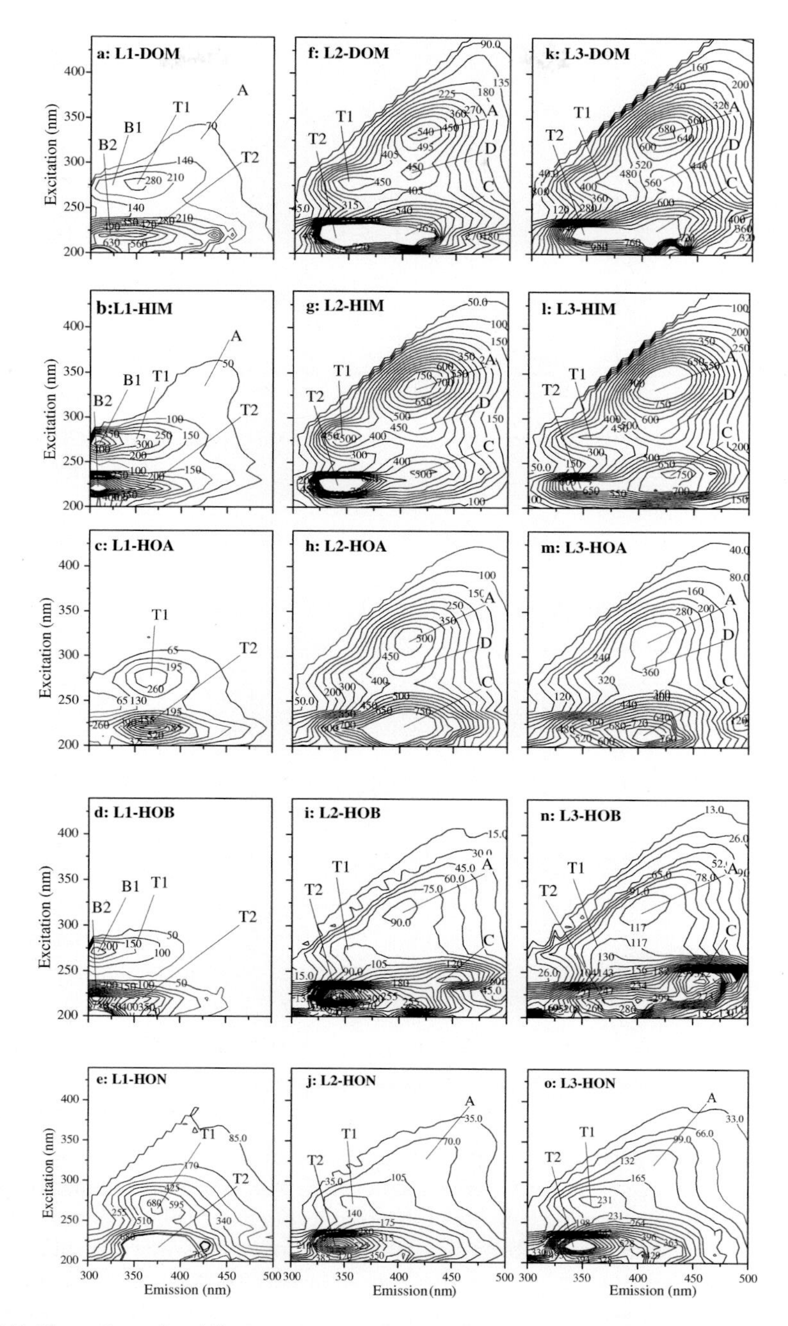

Fig. 2.5 Three-dimensional fluorescence spectrograms of DOM and its components in leachate by landfill ages "Reprinted from He et al. (2011a), with permission from Elsevier"

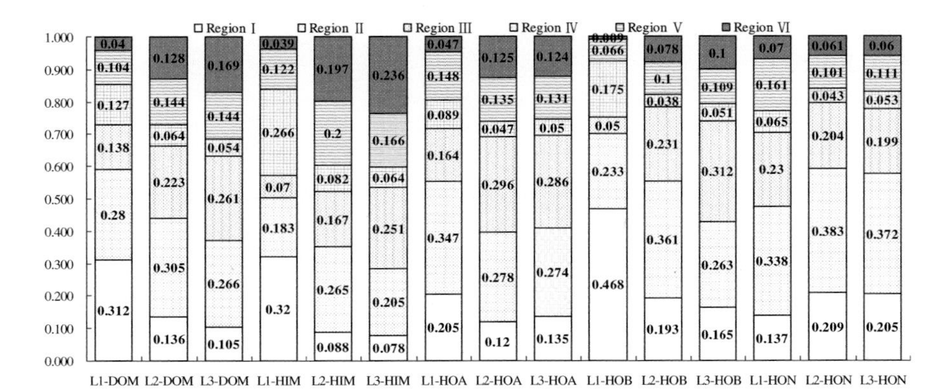

Fig. 2.6 Change in the DOM composition over time "Reprinted from He et al. (2011b), with permission from Elsevier."

Table 2.1 Content of different components and formation constants for characteristic peaks and characteristic fluorescence

	Component	Content ratio (%)	Formation constant			
			Peak T1	Peak T2	Peak C	Peak A
<3 years	HOA	52.25	5.27	5.20	4.20	4.31
	HON	6.96	5.60	5.38	5.48	5.20
	HIM	39.12	5.63	5.46	4.82	5.40
3–10 years	HOA	55.86	4.66	4.59	4.22	4.35
	HON	5.86	5.29	4.93	5.08	5.42
	HIM	37.39	5.24	4.95	4.79	4.99
>10 years	HOA	61.59	5.27	5.13	4.65	4.56
	HON	5.80	5.38	4.71	5.48	4.90
	HIM	31.16	5.76	5.50	4.13	3.88
	Average		5.34	5.09	4.76	4.78

2.3 Technical Methods for Dynamic Multi-objective Optimization of Solid Waste Management

The complex components and large uncertainty of solid waste makes is difficult to weigh the system costs and environmental benefits, posing a severe challenge to the stability and socioeconomic benefits of disposal techniques. This section constructs a model for dynamic multi-objective optimization of solid waste management in uncertain environments based on the considerations of pollution loss, interval uncertainty, infinite programming, and chance-constrained two-stage (CCTS) programming. An optimization system for multi-attribute decision making is built, mitigating the effect of subjectivity and asymmetric information on decisions. On the basis, an optimization-validation-feedback management mechanism that cuts the system costs by 20 % is set forth, conducive to an effective balance of environmental and economic benefits.

2.3.1 *Construction of the Model for Dynamic Multi-objective Optimization Under Uncertainty*

In the context of China's municipal solid waste (MSW) management, considering multiple objectives, uncertainty and constraints, a model for dynamic multi-objective optimization is proposed on the basis of multi-objective planning, pollution loss theory, and uncertainty analysis. The model well integrates economic and environmental aspects and facilitates an objective solution to optimal management of municipal solid waste (Su et al. 2007).

2.3.1.1 General Form of Formula and Solution

The formula of uncertain linear programming can be expressed as follows:

$$Min\ \ f^{\pm} = C^{\pm}X^{\pm} \tag{2.3.1a}$$

$$s.t.\ \ A^{\pm}X^{\pm} \leq B^{\pm} \tag{2.3.1b}$$

$$X^{\pm} \geq 0 \tag{2.3.1c}$$

where, $X^{\pm} \in \{\Re^{\pm}\}^{n\times 1}, C^{\pm} \in \{\Re^{\pm}\}^{1\times n}, A^{\pm} \in \{\Re^{\pm}\}^{m\times n}, B^{\pm} \in \{\Re^{\pm}\}^{m\times l}$; $\Re^{\pm}$ represents a collection of uncertain numbers.

2.3.1.2 Solution of Formula

In the above-mentioned uncertain linear programming formula, the solution necessitates a thorough analysis of the relationship between parameters and variables and between the objective function and constraints. According to Huang et al., an interactive two-step approach can be used: (a) construct and solve the sub-formula for the lower objective function limit f^- (for MIN) and (b) construct and solve the sub-formula for the upper objective function limit f^+, obtaining the uncertain solution of formula. Here is the specific solving process:

In n uncertainty factors of objective function, $cj \pm$ ($j = 1, 2, \ldots, N$), it is assumed that there are k_1 positive numbers and k_2 negative numbers. In other words, the first k_1 factors are positive, i.e. $cj \pm \geq 0$ ($j = 1, 2, \ldots, k1$), and the late k_2 are negative, i.e. $cj \pm < 0$ ($j = k_1 + 1, k_1 + 2, \ldots, n$) and $k_1 + k_2 = n$ (the case of different symbols for lower and upper limits is not included). The algorithm for solving uncertain linear programming problems is as follows:

In the framework of formulas 2.3.1a, b, c the sub-formula for the lower objective function limit f^- can be constructed as follows, (assuming $b_i \pm > 0$):

$$Min f^- = \sum_{j-1}^{k_1} c_j^- x_j^- + \sum_{j-k_1+1}^{n} c_j^- x_j^+ \tag{2.3.2a}$$

$$s.t. \sum_{j=1}^{k_1} |a_{ij}|^+ Sign(a_{ij}^+) x_j^- \Big/ b_i^- + \sum_{j=k_1+1}^{n} |a_{ij}|^- Sign(a_{ij}^-) x_j^+ \Big/ b_i^+ \le 1, \forall i \tag{2.3.2b}$$

$$x_j^\pm \ge 0, j = 1, 2, \ldots, n \tag{2.3.2c}$$

The sub-formula for the upper objective function limit f^+ is built on solution of formulas 2.3.2a, b, c, $x_{j\,opt}^-$ (j = 1, 2,..., K_1) and $x_{j\,opt}^+$ (j = k_1 + 1, k_2 + 2, ..., n), (assuming $b_i \pm > 0$):

$$Min f^+ = \sum_{j-1}^{k_1} c_j^+ x_j^+ + \sum_{j-k_1+1}^{n} c_j^+ x_j^- \tag{2.3.3a}$$

$$s.t. \sum_{j=1}^{k_1} |a_{ij}|^- Sign(a_{ij}^-) x_j^+ \Big/ b_i^+ + \sum_{j=k_1+1}^{n} |a_{ij}|^+ Sign(a_{ij}^+) x_j^- \Big/ b_i^- \le 1, \forall i \tag{2.3.3b}$$

$$x_j^\pm \ge 0, j = 1, 2, \ldots, n \tag{2.3.3c}$$

$$x_j^+ \ge x_{j\,opt}^-, j = 1, 2, \ldots, k_1 \tag{2.3.3d}$$

$$x_j^- \le x_{j\,opt}^+, j = k_1 + 1, k_1 + 2, \ldots, n \tag{2.3.3e}$$

If the objective function is MAX (which requires maximization), the constructing and solving process is contrary to the above. Formulas (2.3.2a, b, c) and (2.3.3a, b, c, d, e) can be adapted to common single-objective linear programming problems. The optimal solution of the formulas (2.3.2a, b, c) f_{opt1}^- can be obtained, i.e. $x_{j\,opt}^-(j = 1, 2, \ldots, K_1)$ and $x_{j\,opt}^+(j = k_1 + 1, k_2 + 2, \ldots, n)$, and the optimal solution of the formulas (2.3.3a, b, c, d, e) f_{opt1}^+, i.e. $x_{j\,opt}^+(j = 1, 2, \ldots, K_1)$ and $x_{j\,opt}^-(j = k_1 + 1, k_2 + 2, \ldots, n)$. Based on this, the final solution is concluded as $f_{opt1}^\pm = [f_{opt1}^-, f_{opt1}^+]$ and $x_{j\,opt}^\pm = [x_{j\,opt}^-, x_{j\,opt}^+]$.

2.3.2 *Model for Dynamic Multi-objective Optimization Under Uncertainty*

2.3.2.1 Principles of Optimization

(1) Reasonable layout and operability

The model is suitable for optimizing the management of solid waste disposal in large and medium-sized cities. As the management involves a large scope and a variety of factors, consideration should be given to urban development and optimization model before ultimately determining a rational layout scheme.

Optimization results rest on implicit assumption about ideal conditions. On the one hand, the results of optimization model should be taken as the scientific basis to maximize economic and environmental benefits. On the other hand, the modeling results should be appropriately adjusted according to actual conditions and economic factors to conform to the comprehensive economic and environmental needs (Su et al. 2007).

(2) Coordination with regional economic development and planning

An optimization model is a phased scientific planning that forecasts the ways and means for solid waste disposal and management in case cities in the next 15–20 years. It encompasses the near-, medium- and long-term development planning and special planning for solid waste treatment and disposal. Hence, an optimization model should take full account urban planning objectives and programs to keep consistency and improve the operability.

(3) Advanced and acceptable technologies

In light of long planned period and accelerating technological progress, it is important to keep abreast of the international solid waste disposal technologies and management methods. The selection of most suitable waste disposal approaches should take into account the advancement and usefulness of technologies while drawing on the experience of the developed countries and advanced domestic practice.

Putting social and environmental benefits in the first place, a forward-looking view that considers the needs of future development is also necessary to form objectives and programs of solid waste management in different stages. The programs should ensure the optimal use of capital, technical and land resources as well as facilities, in order to achieve resourcization and sustainability of solid waste disposal.

(4) Integrity and regional sharing

The urban area is expanding with the accelerated pace of urbanization. Land, capital and technical resources should be configured at the regional level to achieve regional sharing of facilities and resources. According to the systems theory and the synergetic theory, focus should be put on solid waste disposal planning and the effects. The application of optimization models, with a view to minimize the system cost and environmental impact, is expected to break the previous administrative boundaries of cities and facilitate the systemic planning and optimization management of MSW disposal.

2.3.2.2 Planning Scope and Objectives

Principles mentioned above ensure the effectiveness and operability of optimization. The geographical scope of planning should cover the municipal domain of cities and can refer to the development planning for city system, taking into account the suburbs and suburban areas.

In terms of time span, near-, medium- and long-term development planning can be formulated, according to the development plans of cities. To keep in line with the plans for national economic and social development, the short and medium term is generally considered as 5 years and the long term, 20 years (Wan et al. 2005).

The optimization model covers the whole process of waste management, including generation, collection, transportation and disposal. Generally, landfill, composting, and incineration are the major methods for waste disposal in China. Incineration and composting residues are delivered to landfills, so landfill is considered as the final disposal of waste. Economic benefits can be obtained by selling composting products. The whole process is as shown in Fig. 2.7.

According to the overall goals of regional economic and social development and the basic principles of optimal planning, the targets for MSW minimization, recycling and harmless treatment should be set out by scope and stage, including trash removal area and removal rate, separate collection rate and recycling rates, harmless treatment rate, and volume of landfills, as well as levels of mechanization, closure, and modernization.

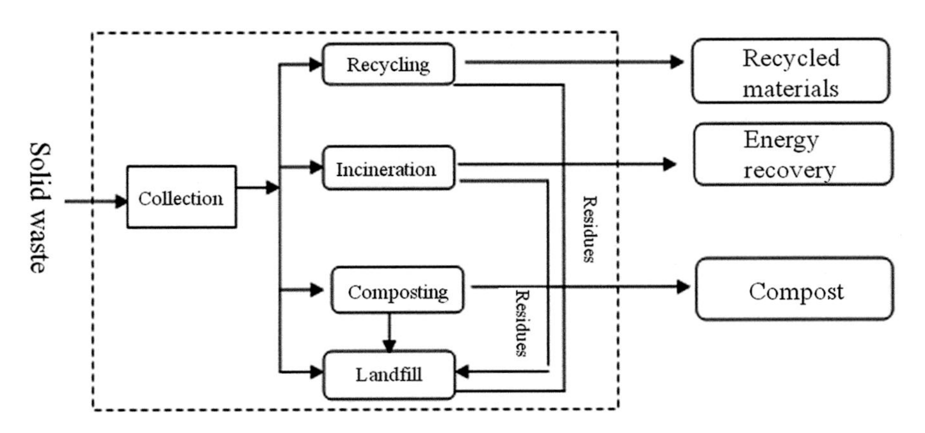

Fig. 2.7 Whole process of solid waste optimization management (Su 2007)

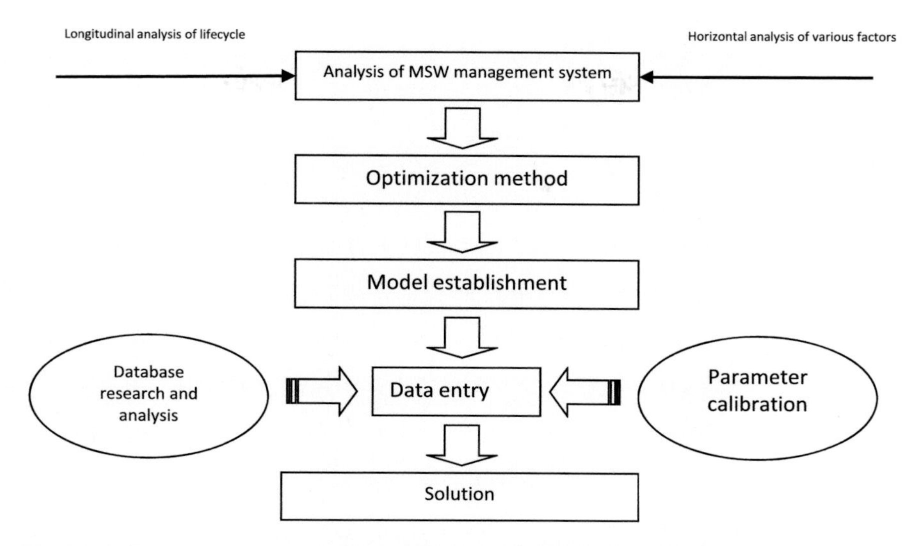

Fig. 2.8 Procedure for building an optimization model for MSW management (Su 2007)

2.3.2.3 Procedure for Building an Optimization Model

The MSW management system is very complex, comprehensive, open, dynamic and uncertain (Zhang et al. 2014). Considering these characteristics, it is suggested to (1) construct an optimization model reflecting regional characteristics based on the scientific analysis of the management system and research of various model; (2) build a modeling framework and conduct database research and analysis and parameter calibration, to form a complete optimization model for MSW management; and (3) in accordance with the solution method, input parameters and obtain the optimal solution (Fig. 2.8).

2.3.2.4 Optimization Model

(1) Objective function

The model is designed to minimize system costs and environmental impact. The constraints include facilities capacity, waste disposal needs, mass balance, ash volume constraints, landfill capacity constraints, pollutant emissions and non-negative constraints. The system costs cover transportation costs, transit fees, processing fees, and material and energy recovery income. The environmental impact is embodied in the costs to reach the national emission standards in the control of secondary pollution from composting, incineration, and landfill (Xi et al. 2007). According to the mentioned objectives and constraints, the function for optimization model is expressed as:

$$
\begin{aligned}
MinZ1^{\pm} = & \sum_{o}\sum_{r}\sum_{t}\frac{(DIS_{ort}\times CYF_{t}^{\pm}+CZF_{t}^{\pm}+CRF_{ort}^{\pm})X_{ort}^{\pm}}{(1+q)^{t}} \\
& +\sum_{o}\sum_{l}\sum_{t}\frac{(DIS_{olt}\times CYF_{t}^{\pm}+CZF_{t}^{\pm}+CLF_{olt}^{\pm})X_{olt}^{\pm}}{(1+q)^{t}} \\
& +\sum_{o}\sum_{i}\sum_{t}\frac{(DIS_{oit}\times CYF_{t}^{\pm}+CZF_{t}^{\pm}+CIF_{oit}^{\pm})X_{oit}^{\pm}}{(1+q)^{t}} \\
& +\sum_{o}\sum_{c}\sum_{t}\frac{(DIS_{oct}\times CYF_{t}^{\pm}+CZF_{t}^{\pm}+CCF_{oct}^{\pm})X_{oct}^{\pm}}{(1+q)^{t}}
\end{aligned} \tag{2.3.4a}
$$

$$
\begin{aligned}
MinZ2^{\pm} = & \sum_{o}\sum_{l}\sum_{t}\frac{(QSHE^{\pm}\times CHEN^{\pm}+QTAN^{\pm}\times CTMQ^{\pm})\times X_{olt}^{\pm}}{(1+q)^{t}} \\
& +\sum_{o}\sum_{i}\sum_{t}\frac{(QFEN^{\pm}\times CFEN^{\pm}+QFEI^{\pm}\times CFEI^{\pm})\times X_{oit}^{\pm}}{(1+q)^{t}} \\
& +\sum_{o}\sum_{c}\sum_{t}\frac{QDUI^{\pm}\times CDUI^{\pm}\times X_{oct}^{\pm}}{(1+q)^{t}}
\end{aligned} \tag{2.3.4b}
$$

where in Z1 represents system costs objective and Z2 environmental impact objective; and ± indicates the maximum and minimum values. *O* stands for the origin of waste, including transfer stations and various treatment facilities; t represents time (years), *r*, recycle stations, *l*, landfill, *i*: incineration plants and *c*, composting plants. DIS stands for distance of transportation (km), *q*, discount rate, and $CYF^{\pm}$ waste transport costs per unit of distance and weight (¥/km, t). $X^{\pm}$ describes the weight of waste (t/d), $CZF^{\pm}$, transfer costs (¥/t), and $CRF^{\pm}$, $CLF^{\pm}$, and $CIF^{\pm}$ indicate unit costs for recycling, landfill and incineration respectively. $QSHE^{\pm}$ means the quantity of leachate produced per unit of waste (t/t) and $CHEN^{\pm}$, leachate treatment costs. $QTAN^{\pm}$ means the quantity of landfill gas produced per unit of waste (t/t) and $CTMQ^{\pm}$, landfill gas treatment costs (¥/t). $QFEN^{\pm}$ means the quantity of exhaust gas produced from incineration per unit of waste (t/t) and $CFEN^{\pm}$, exhaust gas treatment costs (¥/t). $QDUI^{\pm}$ means the quantity of odor produced from composting per unit of waste (t/t); $CDUI^{\pm}$, odor treatment costs (¥/t). $QFEI^{\pm}$ means the quantity of ash generated per unit of waste (t/t) and $CFEI^{\pm}$, ash treatment costs (¥/t).

(2) Constraints

(i) Mass balance

a. Total Amount of Solid Waste = Forecasted Amount of Solid Waste Generation:

$$\sum_{r} X_{art}{}^{\pm} + \sum_{l} X_{alt}{}^{\pm} + \sum_{i} X_{ait}{}^{\pm} + \sum_{c} X_{act}{}^{\pm} = Q_t^{\pm} \qquad (2.3.5a)$$

where a represents transfer stations and $Q_t^{\pm}$ forecasted MSW in year t;

b. Total Amount of Solid Waste from Origin of Waste ≥ Forecasted Amount of Solid Waste Generation;

$$\sum_{r} X_{ort}{}^{\pm} + \sum_{l} X_{olt}{}^{\pm} + \sum_{i} X_{oit}{}^{\pm} + \sum_{c} X_{oct}{}^{\pm} \geq Qt^{\pm} \qquad (2.3.5b)$$

c. Amount of Waste Destined from Transfer Stations for Landfill + Amount of Waste Destined from Treatment Stations for Landfill = Total Amount of Waste Destined for Landfill;

$$\sum_{a} X_{alt}{}^{\pm} + \sum_{r} X_{rlt}{}^{\pm} + \sum_{c} X_{clt}{}^{\pm} + \sum_{i} X_{ilt}{}^{\pm} = \sum_{o} X_{olt}{}^{\pm} \qquad (2.3.5c)$$

d. Amount of Waste Destined from Transfer Stations for Incineration Plants + Amount of Waste Destined from Recycle Stations for Incineration Plants = Total Amount of Waste Destined for Incineration Plants;

$$\sum_{a} X_{ait}{}^{\pm} + \sum_{r} X_{rit}{}^{\pm} = \sum_{o} X_{oit}{}^{\pm} \qquad (2.3.5d)$$

e. Waste that has been treated can no longer return to the recycle stations:

$$\sum_{l} X_{lrt}{}^{\pm} = 0 \qquad (2.3.5e)$$

$$\sum_{i} X_{irt}{}^{\pm} = 0 \qquad (2.3.5d)$$

$$\sum_{c} X_{crt}{}^{\pm} = 0 \qquad (2.3.5g)$$

f. Waste that has been treated can no longer return to the composting plants:

$$\sum_{l} X_{lct}^{\pm} = 0 \quad (2.3.5h)$$

$$\sum_{i} X_{ict}^{\pm} = 0 \quad (2.3.5i)$$

$$\sum_{r} X_{rct}^{\pm} = 0 \quad (2.3.5j)$$

(ii) Ash rate

$$\sum_{i} X_{ait}^{\pm} \times \alpha = \sum_{l} X_{ilt}^{\pm} \quad (2.3.6a)$$

$$\sum_{c} X_{ait}^{\pm} \times \beta = \sum_{l} X_{clt}^{\pm} \quad (2.3.6b)$$

$$\sum_{r} X_{art}^{\pm} \times \gamma = \sum_{l} X_{rlt}^{\pm} + \sum_{i} X_{rit}^{\pm} \quad (2.3.6c)$$

where α represents the residue rate in waste incineration plants, β residue rate in composting plants, and γ residue rate in recycle stations.

(iii) Maximum processing capacity

$$\sum_{i} X_{oit}^{\pm} \leq CAXI \quad (2.3.7a)$$

$$\sum_{c} X_{oct}^{\pm} \leq CAXC \quad (2.3.7b)$$

$$\sum_{l} X_{olt}^{\pm} \leq CAXL \quad (2.3.7c)$$

$$\sum_{r} X_{ort}^{\pm} \leq CAXR \quad (2.3.7d)$$

$$\sum_{a} X_{at}^{\pm} \leq CAXA \quad (2.3.7e)$$

where *CAXI*, *CAXC*, *CAXA*, *CAXL*, and *CAXA* refer to the maximum processing capacity (t/d) of incineration plants, composting plants, landfills, and transfer stations respectively.

(iv) Minimum processing capacity

$$\sum_{i} X_{oit}{}^{\pm} \geq CANI \tag{2.3.8a}$$

$$\sum_{c} X_{oct}{}^{\pm} \geq CANC \tag{2.3.8b}$$

$$\sum_{r} X_{ort}{}^{\pm} \geq CANR \tag{2.3.8c}$$

$$\sum_{a} X_{at}{}^{\pm} \geq CANA \tag{2.3.8d}$$

$$\sum_{t} X_{ol}{}^{\pm} \geq CAMQ \tag{2.3.8e}$$

$$\sum_{l} X_{olt}{}^{\pm} \geq CDMT \tag{2.3.8f}$$

where *CANI*, *CANC*, *CANL*, *CANA*, and *CAMQ* refer to the minimum processing capacity (t/d) of incineration plants, composting plants, landfills, transfer stations and recycle stations respectively. Among MSW, such substances as clinkers, bricks, ceramics and residues after treatment must be buried. *CDMT* stands for the minimum amount of landfill (t/d) (Table 2.2).

(v) Usable components

$$\sum_{i} X_{oit}{}^{\pm} \leq \omega_1 \cdot QBI^{\pm} \tag{2.3.9a}$$

$$\sum_{c} X_{oit}{}^{\pm} \geq \omega_2 QBC^{\pm} \tag{2.3.9b}$$

$$\sum_{r} X_{ort}{}^{\pm} \geq \omega_3 QBR^{\pm} \tag{2.3.9c}$$

Table 2.2 Waste components and applicable treatment methods

Treatment facilities	Type of waste
Landfill	All
Composting	Food waste, plants
Incineration	Food waste, plants, paper, plastics, rubber, textiles, wood and bamboo
Recycling	Metal, glass, paper, plastics, rubber, textiles, wood and bamboo

MSW has complicated and diverse components which pose constraints to disposal methods. Improper operations in waste sorting or disposal hinder ideal waste disposal, resulting in utilization factor constraints. In the above formulas, ω_1, ω_2, and ω_3 represent the utilization factor of incineration, composting, and recycling respectively. In China, Guo Guangzhai et al. concluded the utilization factor for composting and recycling is 90 % based on the MSW research to Shanghai.

(vi) Capacity of treatment facilities

Solid waste treatment inevitably produces secondary pollution. In the multi-objective optimization model, the amount of secondary pollution should be no more than the treatment capacity of facilities to meet the emission standards.

$$QSHE^{\pm} \cdot X_{olt}{}^{\pm} \leq TAXS \quad (2.3.10a)$$

$$QFEN^{\pm} \cdot X_{oit}{}^{\pm} \leq TAXE \quad (2.3.10b)$$

$$QDUI^{\pm} \cdot X_{oct}{}^{\pm} \leq TAXD^{\pm} \quad (2.3.10c)$$

$$QFEI^{\pm} \cdot X_{oit}{}^{\pm} \leq TAXI \quad (2.3.10d)$$

where *TAXS, TAXE, TAXD*, and *TAXI* represent daily processing capacity (t/d) of treatment facilities for leachate, exhaust gas from incineration, composting odor, and ash respectively.

(vii) Non-negative constraints

Non-negative means all the values are greater than zero. The amount of waste out from transfer stations and from treatment plants should meet the non-negative requirement. In other words, $X_{oit}{}^{\pm}$, $X_{olt}{}^{\pm}$, $X_{oct}{}^{\pm}$, $X_{ort}{}^{\pm}$, $X_{ait}{}^{\pm}$, $X_{alt}{}^{\pm}$, $X_{act}{}^{\pm}$, $X_{art}{}^{\pm}$, $X_{ilt}{}^{\pm}$, $X_{clt}{}^{\pm}$, $X_{rlt}{}^{\pm}$, $X_{rit}{}^{\pm}$, and $X_{cit}{}^{\pm} \geq 0$.

(3) Basic data collection, parameter identification, and optimization calculation

The results of optimization model are closely related to the authenticity of parameters. Applicable methods for data acquisition include on-site survey, literature review, and expert consultation. The data obtained by various means should be verified to ensure actual effectiveness.

In the calculation for optimization, it is necessary to change the multi-objective uncertain model into single-objective, deterministic model before solving the formula on the basis of uniform unit and consistent symbol. The model is calculated using LINGO, a simple tool to solve linear and nonlinear optimization problems. Particularly, the language integrated with LINGO can easily express problems and the efficient solver can realize fast solution and effective analysis of the results.

References

Beidou X, Jing S, Jiang Y, et al. Optimization model for municipal solid waste management and influence factors of management costs [J]. Environ Pollut Control. 2007;29(8):561–621.

Cai Q. Case study of the transformation of biological treatment technologies for landfill leachate. Tech Equip Environ Pollut Control. 2003;4(12):76–8.

Cao X, Feng O. Technology for leachate treatment at landfills [J]. Chin Test Technol. 2004;1:38–9.

Chen S, Liu J. Molecular distribution of organic matter in landfill leachate and changes in the MBR system [J]. Environ Chem. 2005;24(2):153–7.

He HS, Xi BD, Zhang ZY, et al. Composition, removal, redox, and metal complexation properties of dissolved organic nitrogen in composting leachates [J]. J Hazard Mater. 2015;283:227–233.

He HS, Xi BD, Zhang ZY, et al. Insight into the evolution, redox, and metal binding properties of dissolved organic matter from municipal solid wastes using two-dimensional correlation spectroscopy [J]. Chemosphere. 2014;117:701–707.

Jia X, Li MX, Xi BD, et al. Integration of fermentative biohydrogen with methanogenesis from fruit-vegetable waste using different pre-treatments [J]. Energy Convers Manag. 2014;88:1219–27.

Li F, Yong W, Bai X, et al. Impact of micro-microbial agents on the nutrients of composting garden waste [J]. Chin Agric Sci Bull. 2012;28(7):307–11.

Li MX, Xia TM, Zhu CW, et al. Effect of short-time hydrothermal pretreatment of kitchen waste on biohydrogen production: Fluorescence spectroscopy coupled with parallel factor analysis [J]. Bioresour Technol. 2014;172:382–90.

Su J. Study to the optimization model for municipal solid waste management [D]. Beijing University of Chemical Technology; 2007.

Su J, Beidou X, Xiujin L. Multi-objective optimization model for municipal solid waste management in uncertain environments [J]. Res Environ Sci. 2007;20(1):129–33.

Xiaosong H, Beidou X, Zimin W, et al. Fluorescence excitation–emission matrix spectroscopy with regional integration analysis for characterizing composition and transformation of dissolved organic matter in landfill leachates [J]. J Hazard Mater. 2011;190(6):239–299.

Xiaosong H, Beidou X, Zimin W, et al. Spectroscopic characterization of water extractable organic matter during composting of municipal solid waste [J]. Chemosphere. 2011;182(1):541–548.

Yang Y, Xiangfeng Z, Zhifeng Y, et al. Nitrogen transformation and nitrogen loss of kitchen waste compost [J]. Environ Sci Technol. 2006;29(12):54–6.

Ying L. Leachate treatment technologies and practical examples [M]. China Environmental Science Press; 2008.

Yuling W. Municipal solid waste management system and mathematical programming [D]. Hunan University; 2005.

Zhang YL, Huo SL, Ma CZ, et al. Using Stressor-Response Models to Derive Numeric Nutrient Criteria for Lakes in the Eastern Plain Ecoregion, China [J]. Clean-soil Air Water. 2014;42:1509–17.

Chapter 3
Classified Resourcization of Solid Waste and Process-Wide Control of Secondary Pollution

Abstract In order to achieve effective recycling of solid waste and control the secondary pollution, the characteristics and potential of the solid waste were studied This chapter expounds on technologies for resourcization of solid waste and control of secondary pollution from a systemic and holistic perspective, covering the whole process from waste collection, transportation and mechanical sorting, bioaugmentation and resourcization, control of secondary pollution, and system integration and management optimization.

Keywords Secondary pollution · Bioaugmentation · Resourcization

3.1 Overview

Solid waste, a metabolic product of municipal lives, has the characteristics of resources and pollutants. It is complex, massive, and widely distributed. Solid waste contains large amounts of recyclable material resources that have huge potential for utilization, but also heavy metals, perishable organic matter, and pathogenic microorganisms which, if not properly disposed, will cause serious pollution. Effective resourcization of solid waste and control of secondary pollution has become a scientific and social problem to be addressed in urban development and rapid urbanization.

To this end, three problems should be first addressed. (1) solid waste disposal is a systematic project, but in China, waste classification, collection, transportation, treatment and disposal is separated from the control of secondary pollution. Given backward key technologies and equipment and lack of integrated solutions, it is difficult to realize the structural and integrated application of best available technologies. (2) Resourcization is to recycle resources according to the characteristics of different components. In light of complex components of solid waste, the lack of an efficient collection, transportation, and sorting system results in low resourcization efficiency and poor product quality, making it difficult to promote resourcization. (3) The control of secondary pollution accompanying solid waste

B. Xi et al., *Optimization of Solid Waste Conversion Process and Risk Control of Groundwater Pollution*, SpringerBriefs in Environmental Science,
DOI 10.1007/978-3-662-49462-2_3

disposal, such as leachate, odor and residues, requires high costs and the effects are poor. It is difficult to mitigate the contradiction between resourcization and secondary pollution and to achieve the unity of environmental and economic benefits.

This chapter analyzes in depth the characteristics of solid waste pollution and potential for resourcization, and expounds on technologies for resourcization of solid waste and control of secondary pollution from a systemic and holistic perspective, covering the whole process from waste collection, transportation and mechanical sorting, bioaugmentation and resourcization, control of secondary pollution, and system integration and management optimization. The idea to maximize reuse and recycling and reduce landfill disposal by classifying and sorting waste: biofortification and humification of organic components, pyrolysis and gasification of combustible components, and recycling of other plastics, metals, glass, and paper.

3.2 Technologies for Classified Solid Waste Collection and Transportation, Biological Pretreatment, and Mechanical Sorting

Classified collection and transportation provides an important guarantee for make use of solid waste. In light of complex components, a source-based waste classification system and corresponding collection and transportation facilities should be put in place, which is an important manifestation of MSW management level. This includes the systems for source-based waste sorting and waste removal. Given that the hybrid mode is prevalent for waste collection and transportation in China, the application of available mechanical sorting technologies is particularly necessary to resourcize waste. In specific, organic components should be separated and stabilized; calorific power of combustible components should be recovered; plastics, metal and paper should be recycled. The practice will greatly reduce the amount of landfills and release the potential of waste as resources.

This chapter analyzes the resource potential and pollution characteristics of solid waste by components, and highlights the technologies for collection and transportation, as well as mechanical sorting applicable to high moisture content waste, while taking into account China's national realities.

3.2.1 Precompression

Compression is an important way to cut the cost of collection and transportation. It also lowers the moisture content, typically by 3–10 %, and reduces secondary pollution during transportation. Generally, compression equipment is comprised of head, precompression box, bi-directional cylinder, shutter, weight and displacement detection system, and hydraulic system. Loose waste can be packaged into blocks after five compressions, and eventually pushed into the docking waste containers.

Transfer station is a key link in the transportation process. To address such problems as large working space, high energy consumption, and long transportation ramp, an efficient transfer system featuring "arrival and departure at the same level" is favored, which increases the transfer energy by more than 10 %. For example, a transfer station in southern China has introduced a horizontal compression system and a vertical compression system for underground packing and transportation, to ensure efficient waste transfer (Fig. 3.1).

3.2.2 *Biological Pretreatment and Mechanical Sorting*

Where a sound system is absent for classified solid waste collection and transportation, solid waste, especially garbage, has high moisture content and complex components. The adhesion caused by moisture transfer and the density differences between components become an important restraint on mechanical sorting efficiency. By separating compostable materials, combustible materials, and utilizable materials, the mechanical sorting technology helps improve the quality of compost, calorific value of raw materials of incineration, and effective system throughput.

3.2.2.1 Dewatering

In solid waste biological pretreatment, dewatering is aimed at losing weight and improving waste separation efficiency. Currently, there are two dewatering options

Fig. 3.1 Transfer station

for biological pretreatment. The first is hydrolysis and fermentation under complex or anaerobic conditions. Making use of damage to cell membrane and organic polymer caused by hydrolyzing bacteria, water bound or attached to food waste is freed and leached under gravity. Due to limitations of water reduction mechanism, the amount of moisture reduction is subject to field moisture content of fermented waste and is generally not less than 55 %. Post-treatment waste shows water saturation and little change in the separation capability. Therefore, this method is mainly applicable to the pretreatment for incineration of high moisture content waste.

Aerobic fermentation is also an option which relies on heat produced in biological reaction to simulate water evaporation. Under ventilated conditions, the water-to-gas transformation serves the dewatering purpose. In comparison, this method can significantly facilitate waste separation. The amount of moisture reduction depends only on heat release and the rate of utilization for water evaporation. However, this method works only under aerobic fermentation conditions and is hardly effective for a mixed collection of refuse. It is obvious that each option has its limitations. To dry mixed waste (moisture reduction) and achieve separate collection and treatment, the dewatering technology for aerobic fermentation pretreatment is developed, which centers on the negative pressure aeration and intelligent ventilation system. By means of high-temperature fermentation and evaporation and negative pressure induced ventilation, the moisture contained in waste can be quickly discharged as steam and leachate. In this process, malodorous gas diffusion is also effectively controlled. Using this technique, the moisture content of waste can be reduced to around 45 % in 12–16 days, drastically shortening the pretreatment cycle, and the material weight can be reduced by more than 40 %. Moreover, this technique changes the particle size of organic components and abates the adhesion between materials, thus increasing the efficiency of mechanical sorting machine.

3.2.2.2 Mechanical Sorting

In line with the Chinese characteristics, mechanical sorting technique and machine featuring a four-stage loop has been developed. Waste is delivered through closed belt conveyor to the primary trommel and crushed and separated by bag knives. The materials are further classified into plastics, paper, and rubber in the secondary trammel screen, achieving a separation efficiency of more than 70–80 %. Then, density-based separation device is used to separate soft plastics, organic waste, and ferromagnetic waste according to density and properties. Finally, the drum sieve ensures the effective separation of small diameter plastics, paper and food waste, with a separation efficiency up to 80 %. Those materials without effective separation will be crushed by biaxial crusher and other crushing equipment, and sent to feed inlet of the sorting line again. The outlets of the four-stage screening system are all equipped with spiral winnowing device, so that the selected materials can be delivered to the plastic separation and cleaning system for recycling. Fully using gravity, wind, and inertia, the four-grade loop mechanical separation process significantly improves waste separation efficiency.

Based on this, three magnetic separation devices are integrated into the combined magnetic separation line to separate poisonous and hazardous magnetic pollutants, such as scrap metal, lighters and batteries, which will address the waste of resources and secondary pollution. Ferromagnetic mixture is delivered to magnetic belt conveyor, two-stage magnetic drum sieve and bouncing sorter, to achieve efficient separation and recovery of ferromagnetic materials of different particle size, density and elasticity.

(1) Primary sorting: trash bag breakage and preliminary separation

Plate feeder conveys the original waste to bag-breaking trommel which is equipped with serrated knives at the entrance. Plastic bags containing waste are automatically broken when moving in the trammel, and the scattered waste are sequentially fed to rotary screen classifier, magnetic separator, bouncing sorter, and winnowing machines to achieve preliminary waste sorting. In this way, a part of masonry, metals, batteries, lighters, wood, fabric, shoes, glass, plastic bottles, rubber, and large pieces of plastics are removed, which will be collected and utilized as recyclable resources. This process significantly reduces stress on follow-up processes and improves oxygen supply for biological pretreatment. The removal of batteries and other harmful substances reduces the heavy metal content of biological compost and improves the quality of organic fertilizers.

(2) Secondary sorting: refined separation and magnetic separation

Following the biological pretreatment, the materials are delivered through loader to the front-end hopper of the secondary sorting system and further subsystems to remove remaining metals. Batteries, wood, fabric plastics, paper, masonry and glass are sorted out, and the remaining organic materials enter to the secondary composting process. A secondary sorting system consists of several subsystems, including frequency regulation feeding system, automatic screening system, material change and rebound system, composite magnetic separation system, plastic dry-cleaning machine, glass sorting system, lighter and battery sorting system, wind-ducted feedback system, and on-line monitoring system (Fig. 3.2).

Fig. 3.2 Mechanical sorting facilities for solid waste

3.3 Technologies for Classified Utilization of Solid Waste

The collection, transportation and mechanical sorting of solid waste according to categories and quality significantly improves the conversion efficiency and enhances product quality of resources.

3.3.1 Biofortification Composting

Compost biofortification and humification can be applied to organic components separated in the mechanical sorting process. By strengthening microbial fermentation, the treatment aims to facilitate the conversion of biodegradable organic matter to stable humus, marking the change from pollutants to resources. The products can be returned to the soil to achieve Earth's carbon cycle and improve the organic matter content of soil (Dhal et al. 2013).

The conventional composting processes face indigenous microorganisms antagonistic to inoculums (Li et al. 2006; Song et al. 2014). The organic components separated preliminarily remain complex, which brings uncertainty to the composting process and secondary pollution. These factors restrict the development of composting technologies.

Temperature control and biofortification inoculation are important means to advance the humification of the compost (Ridha et al. 2012). In the biofortification and humification composting process, temperature control can be achieved, relying on self-produced heat and a little of external heat (24-hour continuous thermophilic fermentation at 70 °C). Pathogenic microorganisms will be quickly killed, and heat-resistant composite microorganisms reserved in rebuilding the compost structure. Meanwhile, thermostat biofortification will make inoculated microorganisms advantageous and maximize the degradation and conversion of substrate, advancing the humification process. Compared with traditional composting techniques, biofortification inoculation significantly increases the degree of aromatic condensation, molecular weight and ratio of oxygen-containing functional groups of humic substances (Fig. 3.4a) (Wei et al. 2012). It means that biofortification can accelerate the protein-like degradation in early stages to form amino acids, enhance the release of hydrophilic quinines from lignin and cellulose, and simulate the formation of humic substances and aromatization in late stages (Fig. 3.4b) (Esperanza et al. 2007). The relative rate of humification can be observed by comparing organic matters in different stages (Fig. 3.4c), providing evidence for optimizing inoculation and hastening humification. Practical applications found that compared with traditional processes, the degradation efficiency of cellulose, hemicellulose and lignin were increased by 15.26, 10.35 and 11.13 % respectively, phenol oxidase activity by 1.5 times, and humification rate (humic acid/fulvic acid) by 24.10 % (Fig. 3.3).

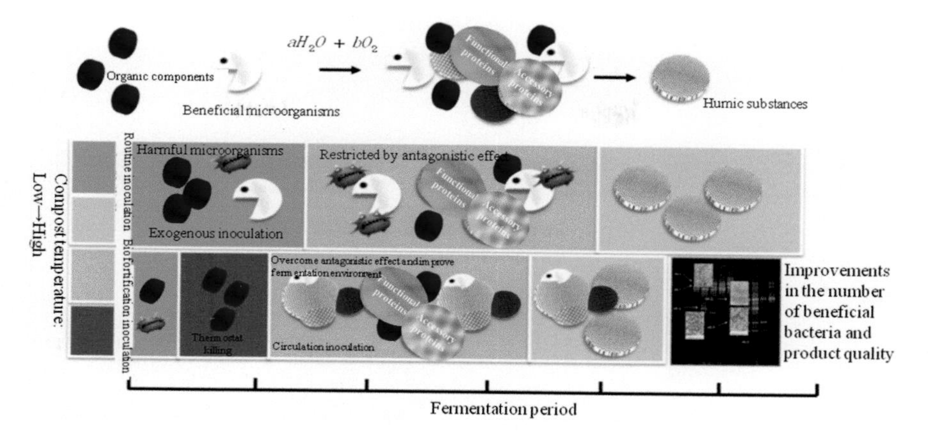

Fig. 3.3 Schematic diagram for thermostat biofortification composting process

Given complex organic components and difficult process control, coupling fuzzy vertex analysis and factor analysis are conducted to study the factors of uncertainty in the materials and process of large-scale biofortification composting. It will help to optimize the control of key parameters, such as substrate, moisture, microbial content and degradation rate, and based on the feedback, better control the composting process and optimize the composting technique. In addition to process-wide control, mechanical standardization is also required so that the links of the process converge efficiently and orderly (Huang et al. 2006) (Fig. 3.5).

3.3.2 *Efficient Anaerobic Digestion*

High contents of water, fat and salt in food waste are not conducive to the fermentation. To this end, hydrothermal pretreatment is introduced, and on the basis, technologies and facilities for high concentration anaerobic fermentation of organic waste mixture are researched and developed, to optimize the solid-liquid coupling parameters and reduce biogas emissions. Further, exploration is made to develop biogas residue fermentation technologies and ecological agriculture that achieve a virtuous circle of organic food waste from land to table and then to land (Li et al. 2014).

(1) Hydrothermal hydrolysis of food waste

Hydrothermal pretreatment is applied to facilitate fermentation of food waste that is complicated by high contents of water, fat, and salt. Studies are conducted to examine the impact of such process parameters as temperature, time, and water proportion on solid-liquid transformation of substances. Hydrothermal hydrolysis can significantly improve the physical and chemical properties of proteins, lipids,

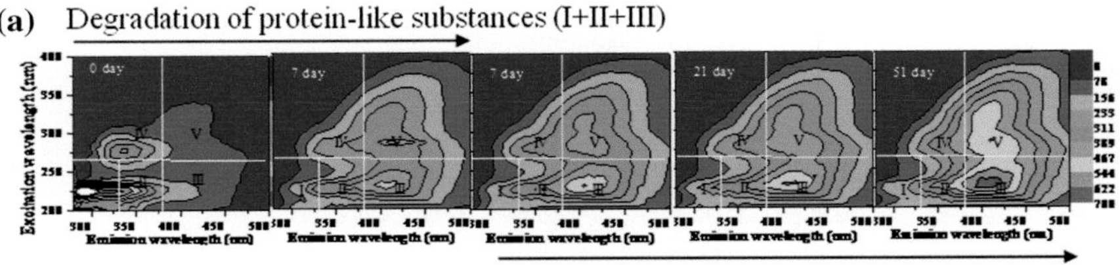

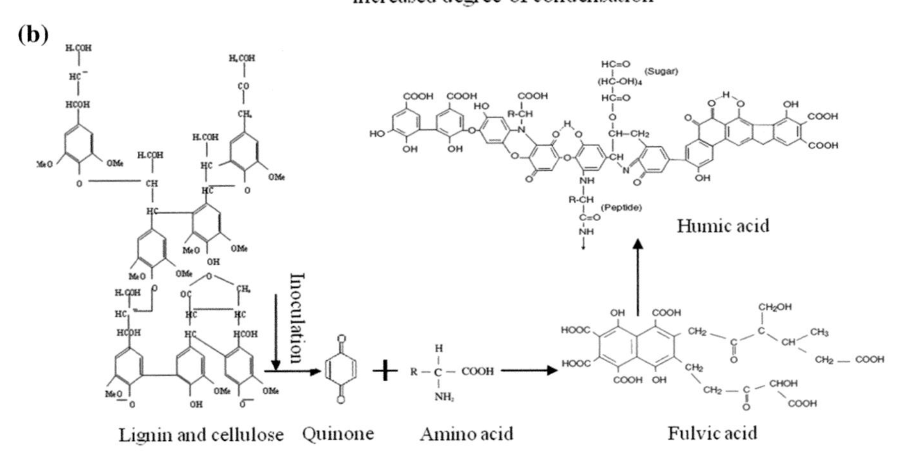

(c) Rescaled Distance Cluster Combine

CASE Label	Num
7 d	2
14 d	3
21 d	4
51 d	5
0 d	1

0 5 10 15 20 25

Fig. 3.4 Diagram for the change of compost structure in the biofortification composting process. **a** Organic matter. **b** Composting process. **c** Relative rate of humification

and carbohydrates, including density, viscosity, ionic product, and dielectric constant, providing an ideal medium for the subsequent anaerobic fermentation. In specific, for proteins, the hydrolysis rate reaches the highest in 70 min at a temperature of 200 °C with 40 % water, up by 44.97 % relative to the control group. Oil slick peaks at 67.7 mL/kg under the conditions of 150 °C, 60 min, and 40 % water, 2.65 times that of the control group. Under the same conditions, the reducing sugar content reaches the maximum of 0.257 g/L, an increase of 16.29 % over the original.

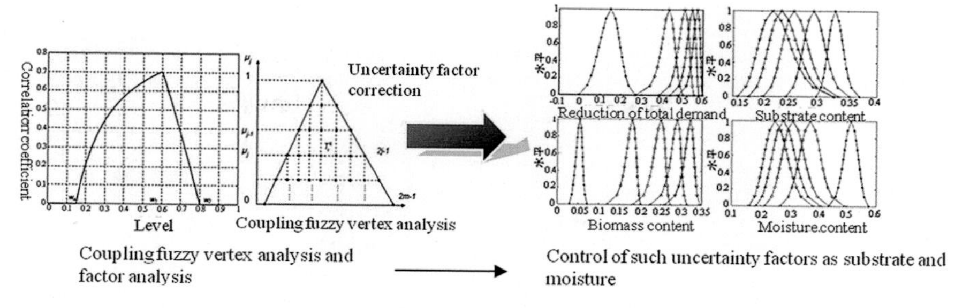

Fig. 3.5 Coupling fuzzy vertex analysis and factor analysis to optimize the composting process

(2) Hydrothermal hydrolysis and anaerobic digestion coupled process

In the anaerobic digestion following hydrothermal hydrolysis, a marked increase is seen in the maximum H_2 production rate of the liquid phase and the maximum CH_4 production rate of mixed materials and the solid phase. According to the results of kinetic curve fitting using the corrected Gompertz, under anaerobic conditions, H_2 production peaks at 546.33 mL in 60 min at a temperature of 150 °C with 40 % water, while the figures reach 1638.93 mL, 50 min, 80 °C, and 60 % for CH_4 production in solid phase. Hydrothermal hydrolysis also increases oil slick, which mitigates LCFAs' microbial inhibition and increases the rate of gas production. Start time is an important indicator of anaerobic fermentation and closely related to the activity of microorganisms and enzyme within the system. The start time of both H_2 and CH_4 production, i.e. the start of anaerobic fermentation, can be delayed by hydrothermal pretreatment. It is because microorganisms are killed during hydrothermal hydrolysis and microbial growth and reproduction is inhibited by the intermediate reductone and volatile heterocyclic compounds generated during pretreatment. Therefore, in the presence of hydrothermal hydrolysis, anaerobic fermentation with solid-liquid separation can achieve the best resourcization of food waste (Table 3.1).

3.3.3 *Pyrolysis and Gasification*

Combustible components mainly include oversize light organics which can be crushed and dried into refuse derived fuel (RDF) for incineration or pyrolysis-gasification. Incineration is widely used and the technologies are mature and stable, particularly in the developed countries. The focus of attention is the environmental risk and human health problems caused by the dioxin emissions. Pyrolysis-gasification is called third-generation waste treatment technology. It was jointly developed by France, the United States, Britain, Germany, Switzerland, Japan and Sweden in the 1990s and has been used in developed countries in the late 1990s (Yuan et al. 2012).

Table 3.1 Analysis of gas production kinetics in anaerobic fermentation

	Conventional CH_4 production	Hydrothermal CH_4 production	Conventional cogeneration	Hydrothermal cogeneration
PH2/mL	4.37	53.23	137.59	105.54
RH2/mL	0.29	3.91	7.89	12.21
PCH4/mL	1658.20	1732.62	2638.16	3519.29
RCH4/mL	16.32	8.03	15.07	15.81
λ/h	282.57	152.50	202.17	175.47
R2	0.99209	0.99159	0.99622	0.99189

The combustible gas produced in pyrolysis-gasification can completely combust in the combustion chamber above 850 °C. The heat can be used by brick kilns and the residue to make building blocks. Following the complete combustion, secondary uniform heat treatment is introduced to completely remove dioxins. The heat can be used by brick kilns; exhaust in the cooling section can be used to dry garbage and the consequent moisture is delivered through specialized equipment. Generally, a large part of moisture is used for RDF gasification and gas combustion and a small part discharged when meeting the EU standards (Figs. 3.6 and 3.7).

3.4 Technologies for the Control of Secondary Pollution

Secondary pollution must be addressed in waste disposal. Resourcization is a way to control secondary pollution from the source, but also it inevitably produces leachate, malodorous gas, and digestate, as well as complex residues with low resource potential and complex pollution characteristics. For this consideration, technologies and facilities that center on resource conversion, waste utilization, and fast stabilization are developed to control secondary pollution.

3.4.1 *In Situ Reduction of Landfill Leachate*

Studies have shown that the leachate recirculation between aged and fresh landfill reactors reduces methane production of methanogenic microorganisms. Although the organic matter reduces with leachate recirculation, the other components (especially ammonia nitrogen) cannot be effectively removed. Therefore, the landfill system fails as the final solution for leachate management (Ke et al. 2004).

The anaerobic-aerobic system is proved viable for in situ simultaneous denitrification. It can effectively remove organic matter and nitrogen from leachate by regulating methanogenesis and denitrification processes. Fresh and aged landfill

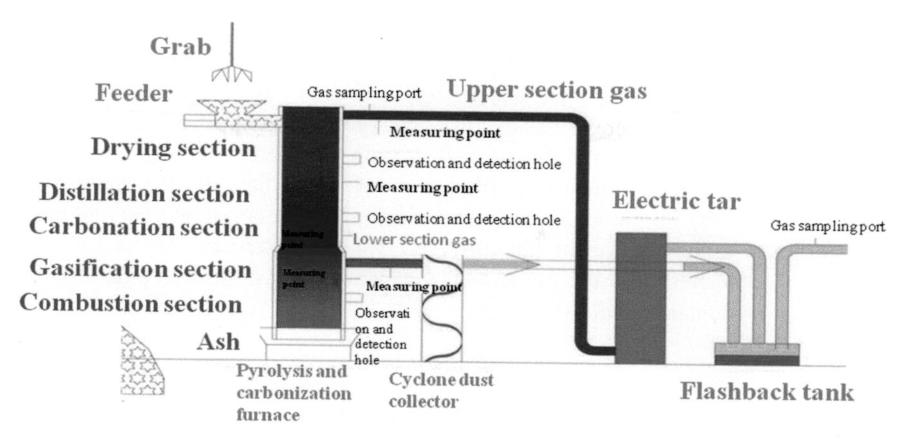

Fig. 3.6 Diagram for gasifier

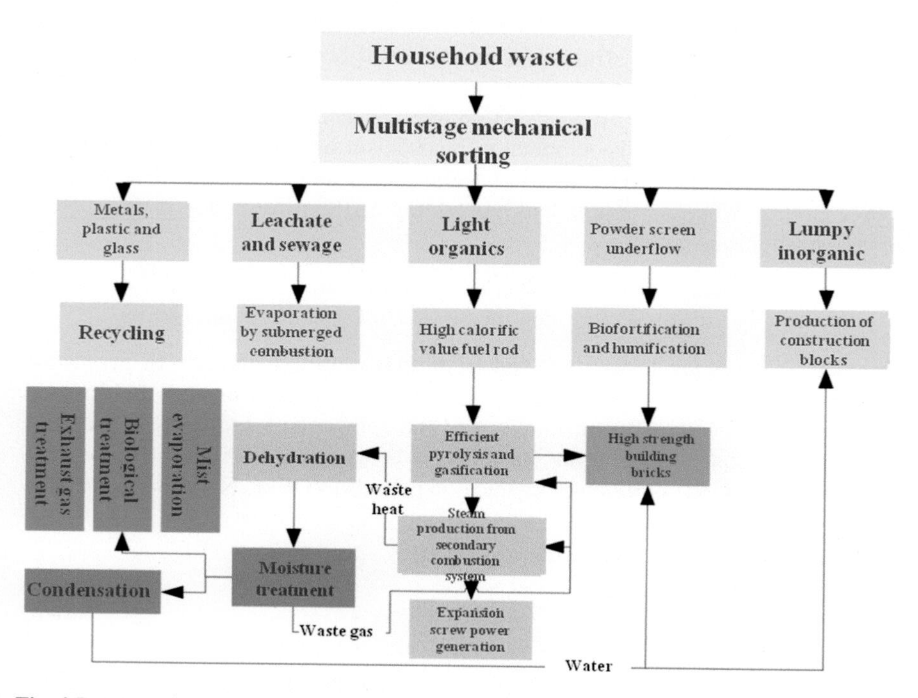

Fig. 3.7 Diagram for gasification process

reactors well support denitrification and methanogenesis respectively, while in the aerobic reactor, ammonia nitrogen can be nitrified.

Recirculated semi-aerobic landfills enable the in situ removal of ammonia nitrogen and organic matter and avoid the off-site treatment of leachate efflux.

In comparison with conventional anaerobic and aerobic bioreactor landfills, semi-aerobic landfills accelerate the degradation of organic matter and nitrogen by providing natural ventilation.

A model is built to observe the dynamic change of biological oxygen demand (BOD_5) in landfill leachate in different modes. With parameter calibration based on experimental data, the simulation results are tested using the efficiency coefficient and T-test methods. The simulation values are found to be consistent with and have no significant differences from the observed values (Fig. 3.8). The sensitivity analysis of parameters lays the foundation for identifying key factors affecting the degradation of contaminants in landfills.

3.4.2 Malodorous Gas Generation and Treatment

Malodorous pollutants are an important part of the secondary pollution of solid waste, which, if not properly controlled, will seriously threaten the ambient air quality and human health. Currently, the technical bottleneck of malodorous gas control lies in the difficulty in controlling the source and grasping the generation and poor capture results. To this end, studies on component changes are made which take into account the production mechanism and absorption and transformation trends. Based on this, a set of technologies and equipment that integrate source control, efficient capture, and biosorption and transformation is designed to control malodorous gases.

(1) Analysis of malodorous gas generation mechanism and component changes in all stages of disposal, and identification of main controllable factors.

Malodorous gases are mainly sourced from anaerobic fermentation of perishable components in household waste, and are usually as complex as household waste. According to the long-term tracking test of the components and content, the most harmful are 8 conventional gases including hydrogen sulfide (H_2S) and ammonia nitrogen and 64 kinds of volatile organic compounds (VOCs). The components and

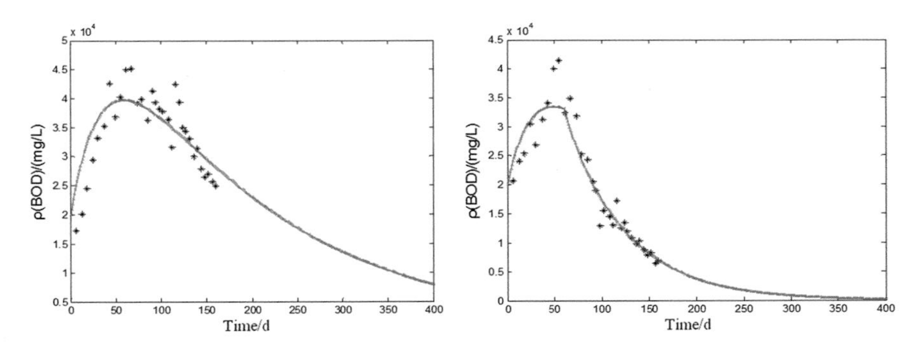

Fig. 3.8 Simulated BOD_5 changes in different landfill modes (Huo et al. 2007)

their content in malodorous gases vary in different stages of disposal, noticeably VOCs. Malodorous gases are emitted mainly in waste transportation and sorting, landfill, and initial fermentation. Currently, the economically feasible and most effective approach is biofilter-based degradation and adsorption, but it needs aid measures to ensure malodorous gas treatment efficiency.

(2) Layered, serial technologies and equipment are developed to effectively control malodorous gases from the humus of organic components of solid waste. This set takes full account of complex effects in the integrated VOCs degradation by humus and the adsorption and degradation characteristics of H_2S and ammonia nitrogen. Using this option, more than 90 % of malodorous gases can be removed.

Bio-fermentation humic substances and adsorptive biomass materials are used as the filler for malodorous gas absorption and removal. In specific, the bio-fermentation humic substances have a high degree of aromatic unsaturation and many carboxyl, carbonyl and other active functional groups, forming strong complex effect on organic pollutants. Meanwhile, while multi-gap structure of filler and microorganism biofilm covering the filler facilitate the adsorption and degradation of such malodorous gases as NH_3 and H_2S. To improve the efficiency of biofilter-based degradation, efficient, layered, serial technologies and equipment are developed. According to the analysis of biodiversity and succession in biofilter at different heights and locations, the top-layer filler with a large area of contact with air presents high degradation rate and sound biodiversity. It is conducive to microbial growth and micro-environmental construction under aerobic conditions. In addition, layering can effectively prevent material compaction caused by accumulation of fillers and the damage to microbial growth environment. Based on this, the outcomes include the layered, serial malodorous gas control technique, which requires a layer thickness of 0.9–1.2 m, a particle size of 4–8 mm for 60 % of fillers, a pH value of 7–8, a temperature range of 20–35 °C, and a humidity range of 45–60 %.

The rule of remove malodorous gases in the composting process can be drawn. NH_3, H_2S, toluene, methyl mercaptan are representative non-volatile gases and VOCs. Four composts of different humification degrees are designed to examine the removal of malodorous substances. The results show that the filler of finished fertilizer presents a high level of gas removal, reaching 98.98 and 98.26 % for NH_3 and H_2S respectively on average. Finished fertilizer reactors perform well in H_2S removal and present average removal rates of 98.93 and 100 % for C_7H_8 and CH_4S.

The rule of biofilter-based approach to remove malodorous pollutants can also be drawn. Finished fertilizer identified based on the laboratory simulation experiments are the optimum filler. A comparative analysis of gas removal effects is made in waste treatment plant in Shanghai before and after applying the filler, covering NH_3, H_2S, carbon disulfide (CS_2), styrene, methyl mercaptan, and dimethyl sulfide. The results show that the filler significantly improves the removal rate. Specifically, the removal rate of methyl mercaptan is increased from 17.22 to 100 %, by replacement 38.10 to 87.95 %, and H_2S from 24.05 to 89.84 %, and CS_2 removal rate is increased by 51 % (Table 3.2).

Table 3.2 Comparative analysis of biolfilter-based gas removal

Gas	Inlet concentration (mg/m^3)		Outlet concentration (mg/m^3)		Outlet rate (kg/h) × 10–5		Removal rate (%)	
	Before	After	Before	After	Before	After	Before	After
NH_3	0.042	0.187	0.026	0.019	1.26	1.83	38.10	89.84
H_2S	0.0158	0.0365	0.012	0.0044	8.85	10.22	24.05	87.95
CS_2	0.727	0.638	0.2804	0.047	35.3	43.58	61.43	92.63
Styrene	0.3178	1.214	0	0	–	–	100	100
Methyl mercaptan	0.00906	0.04262	0.0075	0	90.6	–	17.22	100
Dimethyl sulfide	0.01546	0.03826	0	0	–	–	100	100

Note Dimethyl disulfide and trimethylamine are not included because they are not detected in mechanical sorting and biological pretreatment processes

3.4.3 Biogas Residue Production and Resourcization

Biogas residues are semi-solid substances remained after the anaerobic digestion of solid waste and contain large amount of organic matter, humic acids, crude proteins, nitrogen, phosphorus, potassium and trace elements. In terms of nutrients, humic acids account for 10–20 %, organic matter 30–50 %, total nitrogen 0.8–2.0 %, total phosphorus 0.4–1.2 %, and total potassium 0.6–2.1 %. Given loose texture, good soil moisture performance, and moderate acidity, biogas residues can be converted to fertilizer by composting agents and thereby improve the soil. Biogas slurry, with less than 1 % total solids, contains not only nutrients necessary to crop growth, but also rich amino acids, Family B microbes, plant auxin, and pest inhibition factors, making it an ideal organic fertilizer (Yu et al. 2003).

The research on biogas slurry ecological enrichment with digestate as the main raw materials is carried out. Straw is applied to absorb digestate, and solar power collected by plastic greenhouses and heat produced in aerobic fermentation are combined to accelerate water evaporation. Then, nutrients such as nitrogen, phosphorus and potassium are collected and made into organic fertilizer, and process parameters are thereby determined, including the best formula, nutrient enrichment, and fermentation time.

3.5 Technologies for Integrated Solid Waste Treatment and Management System Optimization

Integrated solid waste treatment is a systematic project that involves all technical links in the resourcization of waste and control of secondary pollution. Process-wide optimization management is needed to achieve best solid waste treatment, lowest management costs, and maximum environmental benefits. The outcomes encompass techniques for designing MSW collection and transportation system, product component prediction and control, dynamic multi-objective optimization in uncertain environments, and optimization—validation—feedback mechanism for optimization management. They ensure system integration and optimization management for classified resourcizaiton of solid waste and control of secondary pollution, and provide technical support for technological selection and project design in waste disposal (Huang et al. 2007).

(1) A solid waste system dynamics forecasting model is established and method designed for forecasting solid waste generation and component changes, to probe into the driving factors and regulation principles (Wan et al. 2005).

The quantity and components of waste are forecasted using the system dynamics model, combined with the gray forecasting model, multiple linear regression model, and backward elimination model (Fig. 3.9). Further, the impact of internal factors, natural factors, individual and social factors is examined, and the study finds that the internal factors play a dominant role and the major factors include urban expansion, population growth, urban development, and improving living standards.

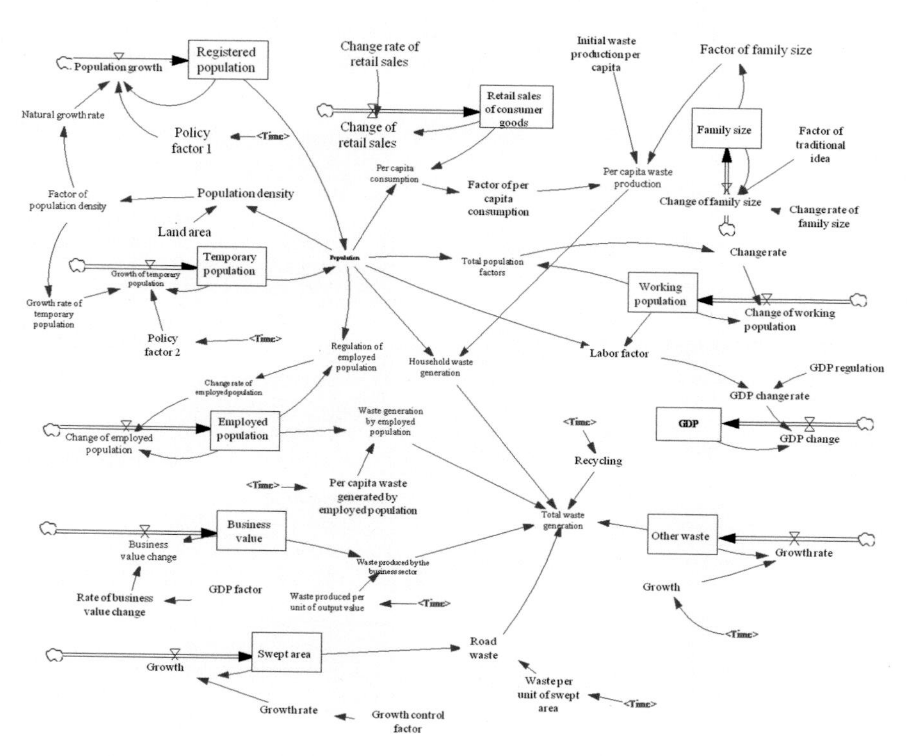

Fig. 3.9 Flowchart for system dynamics forecasting regarding MSW generation (Xi et al. 2010)

Through the case study of Beijing, Shenzhen, and Foshan, the changes of MSW production and components are simulated, and the accuracy rate reaches over 90 %. Meanwhile, the scenario of policy regulation is simulated to observe the changes of MSW production, as well as the organic components, plastics, and metals. The results show that comprehensive control measures can reduce waste generation by 20–30 %, providing a theoretical basis and methodology of MSW disposal project design (Fig. 3.10).

(2) Based on the coupling model for dynamic multi-objective optimization and post-optimization in uncertain environments, techniques are established for system integration and management optimization for classified resourcizaiton of solid waste and control of secondary pollution, providing technical support for technological selection and project design in waste disposal (Su et al. 2007).

Given mixed collection, complex composition and large uncertainty, as well as the challenge in balancing system costs and environmental benefits, dynamic multi-objective optimization in uncertain environments is introduced, and an optimal process combination is proposed, which integrates waste collection and transportation, sorting, recycling/disposal, and the control of secondary pollution. The scheme rests on multi-objective programming, pollution loss theory, interval variables, and parameter uncertainty analysis. It aims to achieve minimum environmental impact and system costs through system design covering the whole process of solid waste treatment and management. Under the scheme, the

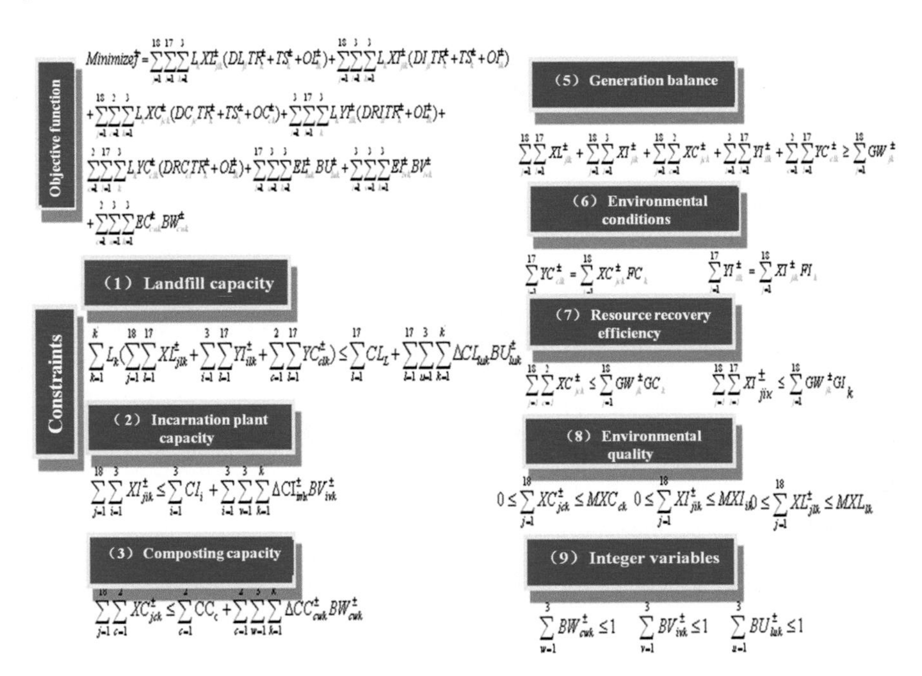

Fig. 3.10 Uncertainty-based dynamic optimization model for solid waste management

constraints include classification, collection and transportation route optimization, process efficiency, sorting rate, resource recovery rate, and secondary pollution emissions. By applying the scheme, the cost was reduced by 28–30 million yuan per year in the case of Foshan.

Taking into account the characteristics of MSW generation and management, multistage optimization is suggested for Beijing, Shenzhen, and Foshan, to improve the disposal model and facilitate management by sections, categories, and stages. In specific, the first stage is mixed collection, mechanical sorting and mixed treatment; the second stage includes crude classification, manual sorting and resource recovery; and the final stage is refined classification and process-wide resource recovery. The multistage optimization will help to achieve efficient resource recovery, harmless disposal, and waste reduction towards sustainable development.

The outcomes combine the chance constrained programming, programming under interval uncertainty, two-stage programming, and multi-attribute decision making system. The resulted optimization-validation-feedback management mechanism improves the technical level of waste disposal and maximizes environmental benefits with minimum economic costs (Fig. 3.11).

Backward tools for data storage, management and analysis have hindered the application of appropriate techniques for MSW treatment and disposal. An

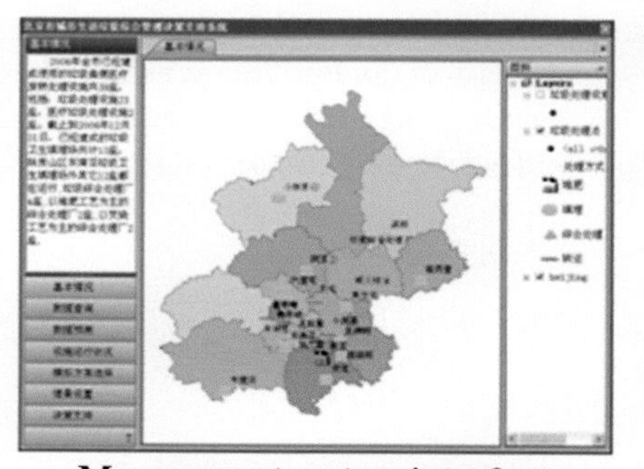

Management system interface

Systemic prediction interface

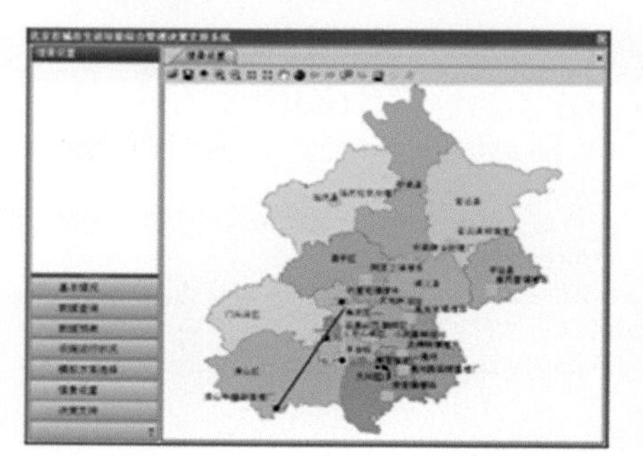

Logistics distribution plan

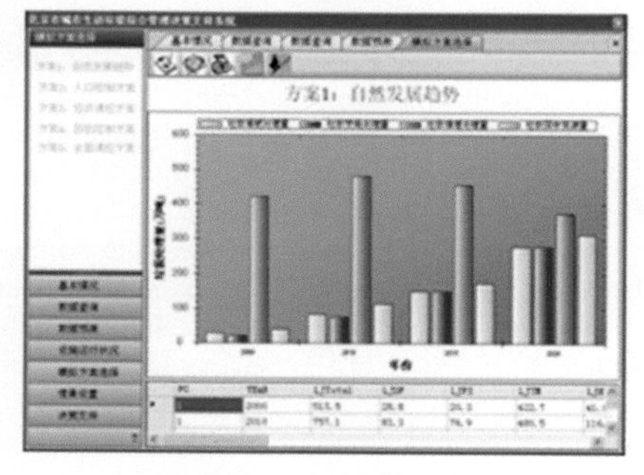

Optimized control system

Fig. 3.11 Model of the decision-making system for integrated MSW management optimization of integrated management techniques is therefore necessary, taking into account the geographic characteristics and economic and environmental benefits under different management models. The optimization will reduce economic costs and environmental pollution, increase recycling revenue, and boost healthy and sustainable development of urban lives. Hence, modern technologies including database and Internet-based information technologies, are applied to improve data management, analysis efficiency, and visualization and operability, providing an effective tool for management decisions. Meanwhile, an integrated management information system is designed, so that experts, policy makers and the public can have access to information, covering solid waste production, collection, transportation and disposal in Beijing's various districts and in different time, as well as recycling and conversion mechanism and system optimization model.

The idea of system design has been embodied in the whole process of solid waste treatment and management. The outcome encompasses a model for dynamic multi-objective optimization in uncertain environments and an optimal process for resource recovery and pollution control. The proposed process integrates waste collection and transportation, sorting, recycling/disposal, and the control of secondary pollution, in order to achieve minimum environmental impact and system costs. The model is built on multi-objective programming, pollution loss theory, interval variables, and parameter uncertainty analysis, and includes constraints, such as classification, collection and transportation route optimization, process efficiency, sorting rate, resource recovery rate, and secondary pollution emissions. On this basis, techniques are established for system integration and management optimization for classified resourcizaiton of solid waste and control of secondary pollution (Xi et al. 2007).

References

Dhal B, Thatoi HN, Das NN, et al. Chemical and microbial remediation of hexavalent chromium from contaminated soil and mining/metallurgical solid waste: a review. J Hazard Mater. 2013;1:272–91.

Esperanza R, Cesar P, Nicola S et al. Humic acid-like fractions in raw and vermicomposted winery and distillery waste. Geoderma. 2007;124:397–406.

Haoran Y, Tao L, Xiong Z, et al. Research progress on pyrolysis and gasification of municipal solid waste. Chem Ind Eng Prog. 2012;31(2):421–7.

Huang GF, Wu QT, Wong JWC et al. Transformation of organic matter during co-composting of pig manure with sawdust. Bioresour Technol. 2006;1834–1842.

Huang J, Li G, Wang H, et al. Evaluation system for solid waste recycling and utilization and research methodology. Environ Pollut Control. 2007;29(1):74–8.

Huo S, Xi B, Fan S, et al. A mathematical model simulating organic variation in bioreactor landfill. Res Environ Sci. 2007;20(5):110–4.

Ke S, Ouyang H. Landfill leachate treatment for municipal solid waste and research progress. Water Wastewater Eng. 2004;30(11):26–32.

Li MX, Xia TM, Zhu CW et al. Effect of short-time hydrothermal pretreatment of kitchen waste on biohydrogen production: fluorescence spectroscopy coupled with parallel factor analysis. Bioresour Technol. 2014;172:382–390.

Li Z. Research on microbial diversity and material transformation in the composting process of agricultural organic solid waste. Nanjing Agricultural University; 2006.

Ridha H, Olfa R, Salma H, et al. Co-composting of spent coffee groun with olive mill wastewater sludge and poultry manure and effect of Trametes versicolor inoculation on the compost maturity[J]. Chemosphere. 2012;88:667–82.

Song CH, Li MX, Jia X. Comparison of bacterial community structure and dynamics during the thermophilic composting of different types of solid wastes: anaerobic digestion residue, pig manure and chicken manure. Microb Biotechnol. 2014;7(5):424–33.

Su J. Study to the optimization model for municipal solid waste management. Beijing University of Chemical Technology; 2007.

Wan Y. Municipal solid waste management system and mathematical programming. Hunan University; 2005.

Wei T, Lingzhi L, Fang L, et al. Assessment of the maturity and biological parameters of compost produced from dairy manure and rice chaff by excitation-emission matrix fluorescence spectroscopy. Bioresour Technol. 2012;110:330–7.

Xi B, Xia X, Su J, et al. Municipal solid waste system analysis and optimization of management technology. Beijing: Science Press; 2010. P. 125–126.

Xi B, Su J, Jing Y, et al. Optimization model for municipal solid waste management and influence factors of management costs. Environ Pollut Control. 2007;29(8):561–621.

Yu H. Life cycle assessment (LCA) method in solid waste resourcization. China Resour Compr Utilization. 2003;09:35–7.

Chapter 4
Solid Waste Disposal and Synergetic Pollution Control

Abstract Bioreactor landfill marks the final disposal of solid waste. Traditionally, mixed waste is buried directly, resulting in low stabilization efficiency. However, the microenvironment of bioreactor landfill is unfavorable for nitrification and blocks nitrogen transformation, leading to nitrogen accumulation and severe leachate pollution. By means of process optimization, in particular, using biofortification, the biofermentation of organic components and landfill of inorganic components are combined to reduce landfills and environmental risks caused by landfill of organic components. To this end, this chapter reveals the rapid reduction mechanism of organic matter, nitrogen conversion process, constraints, and the impact mechanism. Based on this, technical principles and methods for landfill structural optimization and in situ nitrogen reduction are put forward, to effectively decompose organic matter, prevent the nitrogen accumulation, and accelerate landfill stabilization. The practice is crucial to synergies between the disposal of solid waste and control of secondary pollution.

Keywords Bioreactor landfill · Rapid stabilization · Cement kiln · Synergetic pollution control

4.1 Bioreactor Landfill and Synergetic Pollution Control

Bioreactor landfill is considered a final and sustainable method of waste disposal. The combination of biofermentation and landfill can reduce landfills and environmental risks caused by the landfill of organic components.

Bioreactor landfill indicates the biofortification trend of waste disposal technologies. However, the microenvironment in a bioreactor landfill, coupled with complex structure, is unfavorable for nitrification and blocks smooth nitrogen transformation, giving rise to nitrogen accumulation and serious secondary pollution. It is therefore necessary to probe into nitrogen conversion process, constraints, and the impact mechanism, and further put forward the technical principles and

B. Xi et al., *Optimization of Solid Waste Conversion Process and Risk Control of Groundwater Pollution*, SpringerBriefs in Environmental Science,
DOI 10.1007/978-3-662-49462-2_4

methods for landfill structural optimization and in situ nitrogen reduction, so as to effectively decompose organic matter, prevent nitrogen accumulation, and accelerate landfill stabilization.

4.1.1 Organic Matter Reduction Dynamics and Stabilization Technologies

Buried solid waste ultimately stabilizes through a variety of biological, physical, and chemical actions. Landfill stabilization is in fact a process of biochemical reactions characterized by biodegradation of organic matter. In the context of landfill, biodegradation includes a range of microbial activities and biochemical reactions. The stabilization of solid-phase organic matter requires a complex degradation process. Biodegradable organic carbon in solid waste is hydrolyzed and transferred to the liquid phase, while liquid-phase organic carbon is converted by acid-producing microorganisms to CO_2 and organic acids. With acetate as a representative, organic acids are further converted by methanogens to CH_4 and CO_2.

In the early phase of landfill, organic matter degradation is dominant. Protein-like substances and humic-like substances are degraded and aliphatic substituents in the phenyl ring are degraded into carboxyl groups and carbonyl groups, lowering the molecular weight of organic matter. In the middle and late phases, organic matter degradation slows down while humification strengthens. The molecular weight of organic matter and the degree of humification increase over time, improving landfill stability.

Generally, the stabilization of organic matter involves three processes: (1) degradation and disappearance of easily degradable, active organic matter, such as lipids and carbohydrates; (2) synthesis of humic substances accompanied by increase of humus aromaticity, humification degree, and molecular weight; and (3) imbalance in the ratio of microbial nutrients in landfill.

The landfill stabilization process can be tracked by observing the DOM content and spectroscopic properties. When a landfill stabilizes, the DOM content falls below 3.502 g/kg; the DOC-DON ratio is less than 3.693; the ultraviolet (UV) absorbance for DOM below 254 nm is higher than 2.700 L/mg * m; the ratio of characteristic IF absorption values between carboxyl and benzene functional groups is less than 1.840 (Table 4.1). The characteristic fluorescence peaks appear in 280/420 nm 3-D fluorescence spectra (Fig. 4.1).

According to first-order kinetics equation, the conversion (hydrolysis) of organic matter in the bioreactor from the solid phase to the liquid phase can be expressed as follows:

$$\frac{\mathrm{d}S_{si}}{\mathrm{d}t} = -K_{\mathrm{hi}}S_{\mathrm{si}} \quad (i = 1, 2, 3) \tag{4.1.1}$$

Table 4.1 Feature parameters for DOM stabilization in landfills (1995–2003)

Landfill depth (m)	DOC[a] ($g \cdot kg^{-1}$)	DOC/DON[b]	$SUVA_{254}$[c] ($L \cdot mg^{-1} \cdot m^{-1}$)	1635/1406[d]	I_{peak3}/I_{peak1}[e]
0–2	0.476	0.388	2.914	0.979	1.501
2–4	3.502	1.848	4.629	1.275	1.590
4–6	0.544	3.693	3.129	1.118	2.827
6–8	0.843	3.651	4.071	1.010	2.971
8–10	0.383	0.516	2.700	1.841	1.625
10–12	0.703	0.660	3.771	1.433	1.266
12–14	0.564	3.481	3.900	1.824	2.235

[a]DOM concentration; [b]DOC-DON ratio; [c]absorbance value for unit DOM concentration below 254 nm; [d]ratio of absorbance in 1635 and 1406 cm^{-1} in the IR spectra; [e]fluorescence peak ratio between humic-like and fulvic-like substances in 3-D fluorescence spectra

where S_{si} refers to the concentration of component i (organic carbon content, mg/L); (1, 2, 3) represent easily degradable organics, degradable organics, and hardly degradable organics, respectively; K_{hi} stands for the first-order hydrolysis constant for component i (d^{-1}).

The conservation from liquid-phase organic matter to acetic acid and from acetic acid to CH_4 can be expressed by Formulas (4.1.3) and (4.1.2), respectively.

$$\frac{dS_A}{dt} = -K_A S_A \tag{4.1.2}$$

$$\frac{dS_M}{dt} = -K_M S_M \tag{4.1.3}$$

where K_A indicates the average rate constant of acetic acid generation (mg/L) and K_M, methane (mg/L).

Microbial growth is expressed by the Monod equation:

$$\mu = \frac{\mu_M S}{K_s + S} \tag{4.1.4}$$

where μ represents the microbial growth rate $\mu = \frac{1}{X} \cdot \left(\frac{dx}{dt}\right)$ (mg/Ld), and μ_M maximum specific growth rate (d^{-1}). S indicates the substrate concentration (mg/L) and X microorganism concentration (mg/L). K_s is the half-saturation constant (mg/L).

Y stands for the amount of microorganism yield per unit of substrate, expressed as follows:

$$Y = -\frac{dX}{dS} \tag{4.1.5}$$

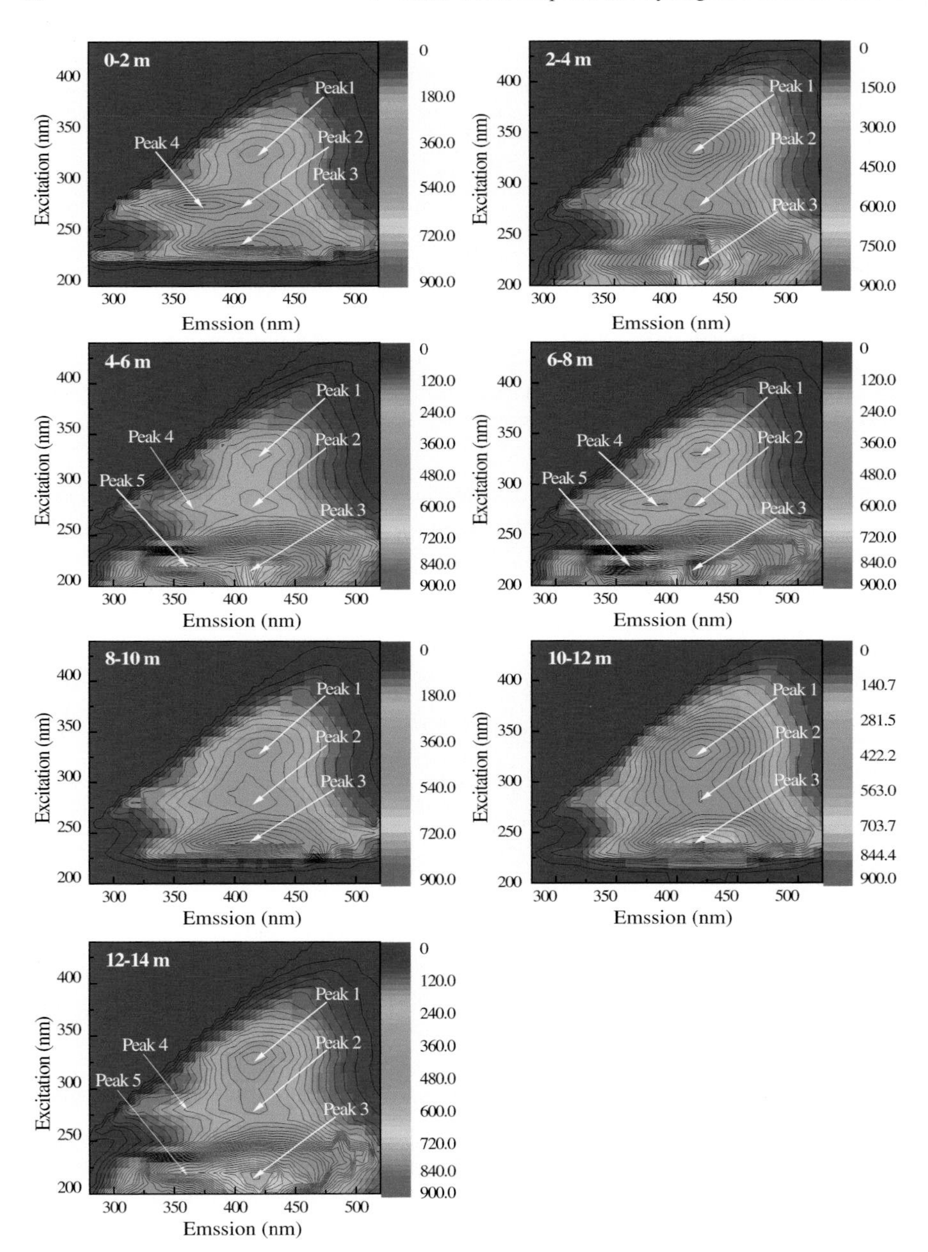

Fig. 4.1 Three-dimensional fluorescence spectrogram for DOM in landfills (1995–2003)

The substrate consumption rate is expressed as follows:

$$\frac{dS}{dt} = -\frac{dX}{Ydt} = -\frac{XdX}{XYdt} = -\frac{\mu X}{Y} = -\frac{\mu_M SX}{Y(K_S + S)} \tag{4.1.6}$$

In anaerobic biodegradation process, substrate is used to support microbial growth and maintain endogenous microbial metabolism. The later part of consumption is expressed by K_d, i.e., constant of microbial mortality rate. The actual growth rate of microorganisms can be drawn as follows:

$$\frac{dX}{dt} = (\mu - K_d)X \tag{4.1.7}$$

The growth rate of acetogenic bacteria is expressed as follows:

$$\frac{dX_A}{dt} = \left[\left(\frac{\mu_A S_{aq}}{K_{SA} + S_{aq}}\right) - K_{dA}\right]X_A \tag{4.1.8}$$

where X_A is the amount of acetogenic bacteria (mg/L) and S_{aq} the DOC concentration (mg/L). K_{SA} is the half-saturation constant for acetogenic bacteria (mg/L), K_{dA} death rate constant (day^{-1}), and μ_A maximum specific growth rate (d^{-1}).

Methanogen growth rate is expressed as follows:

$$\frac{dX_M}{dt} = \left[\left(\frac{\mu_M S_{AC}}{K_{SM} + S_{AC}}\right) - K_{dM}\right]X_M \tag{4.1.9}$$

where X_M is the amount of methanogens (mg/L)and S_{AC} the acetic acid concentration (mg/L). K_{SM} is the half-saturation constant for methanogens (mg/L), K_{dM} death rate constant (day^{-1}), and μ_M maximum specific growth rate (d^{-1}).

In the bioreactor, the DOC concentration is influenced by the hydrolysis of solid-phase organic carbon and the conversion to acetic acid. Under the combined effects, the change rate of DOC concentration is equal to the hydrolysis rate of solid-phase organic carbon minus the rate of conversion to acetic acid. It can be expressed as follows:

(Change Rate of DOM within the Reactor) = (Hydrolysis Rate of Solid-Phase Organic Carbon) − (Rate of Conversion to Acetic Acid)

It is expressed by the following equation:

$$\frac{dS_{aq}}{dt} = \frac{dS_s}{dt} - \frac{dS_A}{dt} \tag{4.1.10}$$

The following equation can be obtained after giving inputs to Formulas (4.1.1) and (4.1.6):

$$\frac{\mathrm{d}S_{aq}}{\mathrm{dt}} = \sum_{i=1}^{3} (K_{\mathrm{hi}} S_{\mathrm{si}}) - \left(\frac{\mu_A S_{aq}}{Y_A (K_{SA} + S_{aq})} \right) X_A \tag{4.1.11}$$

The physical balance of acetic acids involves three portions: acetic acid converted from DOC, generated in the reaction and used by microorganisms. It can be expressed as follows:

$$\frac{V\mathrm{d}S_{AC}}{\mathrm{dt}} = \frac{Y_{\mathrm{HAC}} V\mathrm{d}S_A}{\mathrm{dt}} - \frac{V\mathrm{d}S_M}{\mathrm{dt}} - \frac{Y_{\mathrm{HAC}} V\mathrm{d}X_A}{\mathrm{dt}} \tag{4.1.12}$$

$$\frac{\mathrm{d}S_A}{\mathrm{dt}} = -\frac{\mu_A S_{aq} X_A}{Y_A (K_{SA} + S_{aq})} \tag{4.1.13}$$

$$\frac{\mathrm{d}S_M}{\mathrm{dt}} = -\frac{\mu_M S_{AC} X_M}{Y_M (K_{SA} + S_{AC})} \tag{4.1.14}$$

The following equation can be obtained after giving input to Formulas (4.1.13) and (4.1.14):

$$\begin{aligned} \frac{\mathrm{d}S_{AC}}{\mathrm{dt}} = {} & Y_{\mathrm{HAC}} \left(\frac{\mu_A S_{aq} X_A}{Y_A (K_{SA} + S_{aq})} \right) - Y_{\mathrm{HAC}} \left[\left(\frac{\mu_A S_{aq}}{K_{SA} + S_{aq}} \right) - K_{dA} \right] X_A \\ & - \frac{\mu_M S_{AC} X_M}{Y_M (K_{SM} + S_{AC})} \end{aligned} \tag{4.1.15}$$

It can be rewritten as follows:

$$\begin{aligned} \frac{\mathrm{d}S_{AC}}{\mathrm{dt}} = {} & Y_{\mathrm{HAC}} \left[(1 - Y_A) \frac{\mu_A S_{aq}}{Y_A (K_{SA} + S_{aq})} + K_{dA} \right] X_A \\ & - \left(\frac{\mu_M S_{AC}}{Y_M (K_{SM} + S_{AC})} \right) X_M \end{aligned} \tag{4.1.16}$$

Similarly, the equations for the change rate of CH_4 and CO_2 are written as follows:

$$\frac{\mathrm{d}C_{\mathrm{CH_4}}}{\mathrm{dt}} = Y_{\mathrm{CH_4}} \left[(1 - Y_M) \frac{\mu_M S_{AC}}{Y_M (K_{SM} + S_{AC})} + K_{dM} \right] X_M \tag{4.1.17}$$

$$\begin{aligned} \frac{\mathrm{d}C_{\mathrm{CO_2}}}{\mathrm{dt}} = {} & (1 - Y_{\mathrm{HAC}}) \left[(1 - Y_A) \frac{\mu_A S_{aq}}{Y_A (K_{SA} + S_{aq})} + K_{dA} \right] X_A \\ & + (1 - Y_{\mathrm{CH_4}}) \left[(1 - Y_M) \frac{\mu_M S_{AC}}{Y_M (K_{SM} + S_{AC})} + K_{dM} \right] X_M \end{aligned} \tag{4.1.18}$$

where Y_A and Y_M represents the amount of acetogenic bacteria and methanogens produced per unit of organic matter, respectively (mg/mg). Y_{AC} is the acetic acid production coefficient (kg/kg), taking 0.9 and Y_{CH_4} methane production coefficient, taking 0.6–0.7 (mg/mg). C_{CH_4} stands for the CH_4 concentration (mg/L) and the CO_2 concentration (mg/L).

4.1.2 Nitrogen Conversion Process and In Situ Removal Optimization Mechanism

Indicators such as $SUVA_{254}$ and DOC/DON are used to examine the degree of landfill stability. According to the study on nitrogen conversion process, constraints, and the impact mechanism in recirculated, semi-aerobic, and sequencing batch bioreactors, in an anaerobic environment hindered nitrification, are the key restraints on the successful nitrogen conversion (Fig. 4.2a). To this end, structural optimization methods are set forth, such as semi-aerobic and recirculated combination and intermittent oxygen presence, to create the anaerobic and aerobic microenvironments favorable for nitrification, organic matter degradation, and landfill stabilization. Particularly, key technologies for in situ nitrogen removal optimization are developed, providing an effective solution to the ammonia accumulation problem (Fig. 4.2b).

The research findings were often referred to by Comstock et al. (2010) and Muller (2011). For example, Greek scholar Aris Nikolaou said that "in bioreactors with limited aeration, researchers have shown that the intermittent presence of oxygen in the bioreactor is favorable to denitrify the NO_2^{-N} and NO_3^{-N}, as it was indicated by the high denitrification efficiency".

4.1.3 Interaction of Pollutants in the Leachate and Discontinuous Permeable Reactive Barrier (PRB)

The study combines Resonance Rayleigh scattering and fluorescence quenching with parallel factor analysis. It is found that in the process of organic matter degradation, carboxyl groups, and carbonyl groups on the benzene ring increase, leading to the redshift of DOM maximum fluorescence peaks and advancing the humification process. Based on the humus-protein fluorescence ratio, a method is put forward for fast prediction of leachate biodegradability (Fig. 4.3b). With increasing humification rate and polarity, different components (HOA, HIM and HON) become more able to bind heavy metals mercury and polycyclic aromatic hydrocarbons. It is conducive to the mitigation of bioavailability, toxicity, and

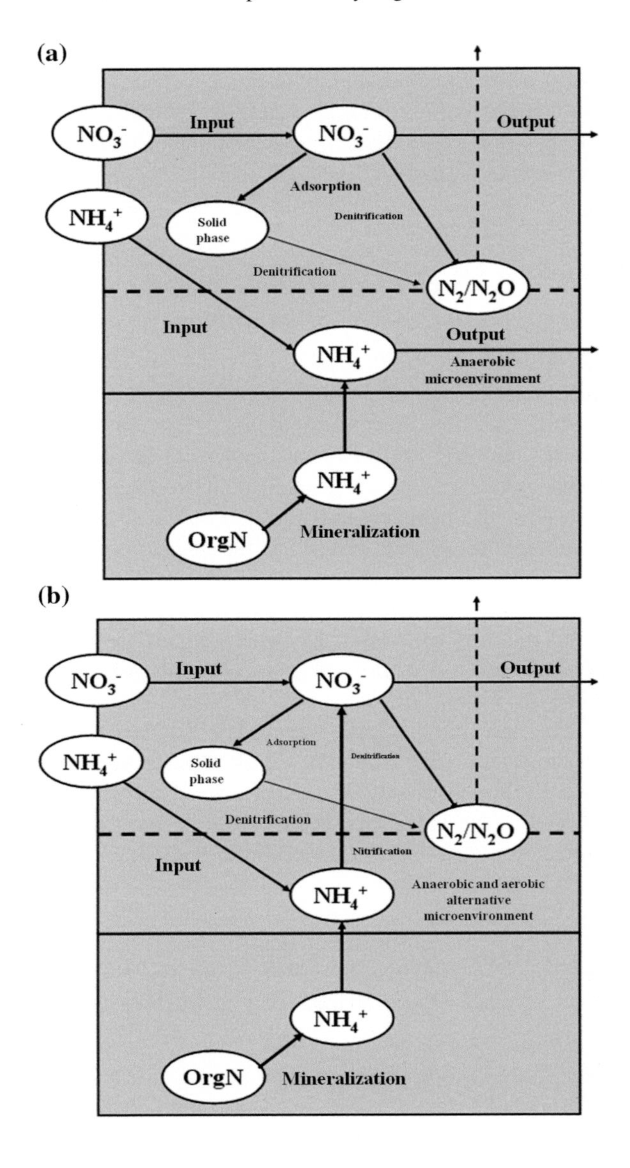

Fig. 4.2 Nitrogen conversion process comparison between different landfills. **a** Nitrogen accumulation process in traditional landfill. **b** In situ nitrogen reduction process in bioreactor landfill

environmental risks of such pollutants as mercury and phenanthrene (Fig. 4.3a). Moreover, groundwater contamination caused by leachate can be diagnosed by identifying the heterogeneous organic matter (Fig. 4.3c), and discontinuous PRB technologies and active media dedicated materials are developed.

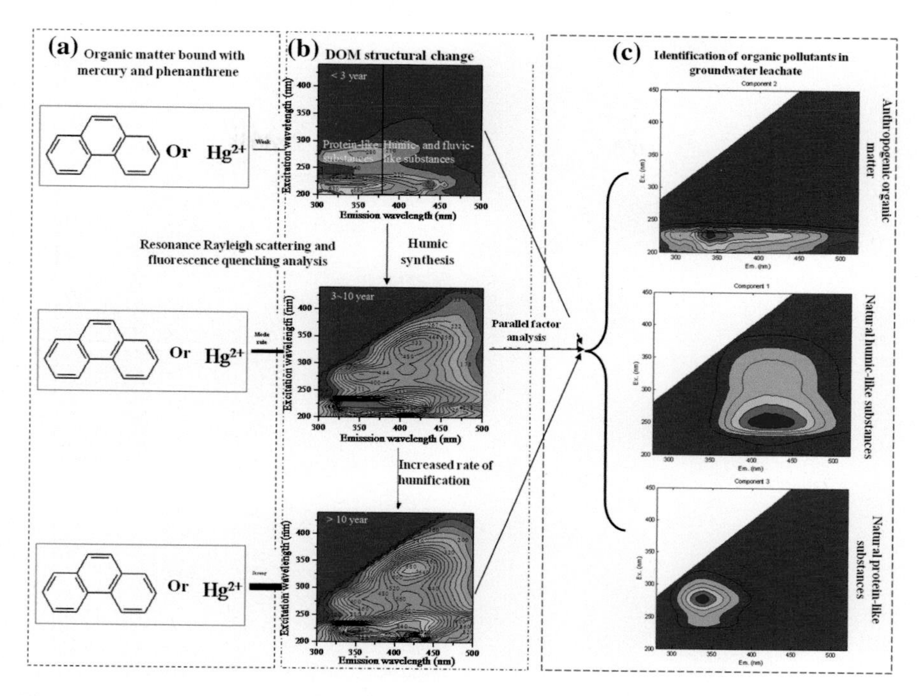

Fig. 4.3 Structural change, environmental effects, and pollutant identification

4.1.4 Engineering Applications for Rapid Stabilization of Bioreactor Landfill and Collaborative Control of Secondary Pollution

In a demonstration project for waste bioaugmentation and disposal (120 tons/day) in Hebei Province, China, the technical route of mechanical sorting, aerobic bioaugmentation, and landfill is adopted. By the way of mechanical sorting, organic and inorganic components are separated from the source, plastics, and metals recycled. Then, organic matter bioaugmentation for fast reduction is introduced to enhance humification and reduce the moisture content of waste and subsequent leachate contamination. This minimizes the risk of environmental pollution while increasing landfill capacity. The stable components including inorganic matter are buried, and the humic products of organic waste can be used as landfill soil and methane oxidation layer materials to reduce CH_4 and other greenhouse gas emissions, landfill costs, and secondary pollution. Generally, the methane adsorption capacity of such soil can reach 10–25 L CH_4/m^2 * h. In terms of landfill construction, the optimization of bioreactor structure and operation mode by design can realize the in situ simultaneous removal of organic matter and ammonia nitrogen

Fig. 4.4 Pictures of project site

from leachate. It will reduce the pollution directly caused by raw waste landfill and achieve the combination of waste bioaugmentation and disposal and ultimately synergies between rapid stabilization of landfill and control of secondary pollution (Fig. 4.4).

4.2 Hazardous Waste Treatment and Synergetic Cement Kiln Treatment

4.2.1 Treatment and Disposal of Typical Hazardous Waste

National Catalogue of Hazardous Wastes includes more than 600 kinds of hazardous waste in 47 categories. In China, hazardous waste generated in the industrial sector amounted to 31.569 million tons (in 2013), of which over 50 % can be comprehensively utilized, including half by disposal (including incineration and landfill) and half by storage. By the end of December 2012, there were 37 sets of hazardous waste facilities (including those put into formal operation and trial operation and basically completed) in 56 hazardous waste disposal projects nationwide, making a total capacity of 1.42 million tons/year. In addition, there were around 1700 enterprises engaging in hazardous waste business (including medical waste) (Hu and Zhang 2014a, b).

4.2.1.1 Fly Ash from MSW Incineration (MSWI)

(1) Fly ash characteristics
MSWI power generation developed in recent 50 years and it is considered as one of the best measures for "resource recovery, harmless treatment, quantity reduction". Giving comprehensive consideration to social, environmental, and economic benefits; the measure is viewed as an optimal option to address the waste siege and environmental pollution. However, the fly ash produced by incineration contains heavy metals and dioxins and requires special treatment (He et al. 2003). In order to achieve harmless treatment, quantity reduction, and resource recovery, a large number of studies have been carried out at home and abroad, but the suggested technical routes are deficient to certain extent, mainly attributed to heavy metals, soluble salts, and dioxins in the fly ash. Currently, the prevalent approach for fly ash disposal is safe landfill. However, in economically developed areas, site selection difficulty coexists with limited storage capacity and high construction costs for new hazardous waste landfill. It is therefore necessary to find new methods of fly ash disposal to make use of fly ash and reduce secondary pollution and long-term adverse effects caused by incineration (Zeng et al. 2012).

(2) Treatment and disposal technologies
To meet the growing demand for MSWI treatment, a lot of research has been made at home and abroad regarding the safe disposal and resourcization of MSWI fly ash. Solidification and stabilization is an internationally prevalent method for the disposal of toxic waste. Fly ash is buried or utilized after harmless treatment by means of cement solidification, melting solidification, or stabilizing agents.

(a) Solidification
Cement solidification is one of the most common techniques for hazardous waste solidification. The purpose is to stabilize such toxic and hazardous substances as fly ash through harmless disposal, and the basic principle is decreasing the surface area and permeability of solidified wastes.
Agent-based stabilization is the process of changing toxic and hazardous substances to substances with low solubility, migration, and toxicity through chemical reactions with the application of chemical agents. These stabilizing agents are water-soluble chelating polymers that react with heavy metal ions to form strong chemical bonds and generate stable complex or insoluble precipitate (Lombardi et al. 1998; Ubbnaco and Calabrese 1998).

(b) High temperature stabilization and recycling

High-temperature stabilization is to turn the fly ash to glassy silicate at high temperature so that the soluble components are packed. The technique can remove such harmful substances as heavy metals and dioxins from fly ash at high temperature, and the residues can be used for construction, achieving the utilization of fly ash. By this means, the volume of residues is significantly reduced and the vitreum is so stable to secure fly ash stabilization. The technique of high-temperature stabilization has been widely used in the United States, Germany, and Japan for harmless treatment of MSWI fly ash for reduction and re-sourcization (Mangialardi et al. 1999).

The sintering process also serves for solidification and stabilization, in which hazardous waste and small vitreous, such as glass frit, are uniformly mixed and granulated by machine or by hand, and vitrified at 1000–1100 °C. Given the dense crystal structure, solidified vitreous can be permanently kept. Currently, melting and solidification is the most advanced method of MSWI fly ash treatment. In essence, fly ash is melted at high temperatures and then cooled to form the desired materials, where the temperature is supported by fuel furnace or electric power, and generally, maintained above 1000 °C. Fly ash—cement kiln integrated treatment is also an important direction of research on high-temperature fly ash at home and abroad.

(c) Extraction and separation

Fly ash extraction and separation involve water washing, acid washing, ion exchange, magnetic separation, microbial separation, and electric separation. This process can: (1) remove and separate specific components and elements from fly ash, such as heavy metals and soluble salts; and (2) improve the engineering properties by classifying fly ash according to the particle size and grade and mixed solution conductivity, which is more conducive to follow-up treatment. Water and acidic solution are most often used for extraction and separation. The method for ash treatment is simple, and in most cases, used together with other methods, such as acid washing and chemical stabilization. It presents superior performance in the recovery of valuable resource and safe disposal of fly ash at low costs. However, a large amount of waste water is produced in the process, requiring further treatment, and fly ash cannot be completely disposed, requiring enhancement by other methods (Quina et al. 2008).

(d) Other approaches

Fly ash contains a large amount of SiO_2, a kind of silicate needed for ceramics production. Given that fly ash is a powder produced in waste incineration, the small particles contained can be used directly as raw materials for ceramic without processing. However, the large presence of metal materials is likely to affect the performance of ceramic, so the application of fly ash must be strictly controlled. Studies showed that it is best to introduce 50 % fly ash. Under the conditions, ceramics have sound physical strength, are not easy to break, and meet the national heavy metal leaching standards and therefore can be used in the actual construction (Bernardo 2007; Rad and Alizadeh 2009).

(3) Problems and trends

Fly ash has a certain value of resources theoretically and the utilization is technically feasible. Nevertheless, prior to 2008, fly ash can only be buried in hazardous waste landfills. In July 2008, the *Standards for the Control of Pollution from Municipal Solid Waste Landfills* came into force stipulating that pretreated fly ash that meet certain conditions can be disposed by landfill. However, in fact, due to the lack of supervision and higher economic costs, fly ash disposal often deviates from formal channels in various forms. In some cases, fly ash is mixed with slag. Compared with hazardous waste landfills, common MSW landfills are less impervious to fly ash, giving rise to leaching in the presence of sewage. In some cases, pretreated fly ash is not buried separately as required after it is transported to sanitary landfills.

The physical and chemical properties, including particle size and element composition, make fly ash potential raw materials for resource products. The research on techniques for fly ash stabilization and resourcization is greatly needed to ultimately achieve waste utilization.

4.2.1.2 Characteristics of MSWI Fly Ash in the Melting Process

(1) Changes in major components

MSWI fly ash contains complex components, of which the content is related to a variety of factors, including waste categories and pretreatment, incinerator, incineration parameters, and flue gas treatment. The major components of fly ash include CaO, SiO_2, Al_2O_3, SO_3, K_2O, Na_2O, and Cl, accounting for about 90 % of the total by weight. It is therefore particularly important to study the variation of these substances in the process of melting and solidification. In the melting process, the contents of CaO, Al_2O_3 , and SiO_2 increase with temperature, and the total percentage in slag approximately grows to 90 % from the original level of 47 % when the temperature rises from 1260 to 1350 °C. In contrast, the contents of SO_3, K_2O, Na_2O, and Cl decrease with rising temperature. At a temperature of 1350 °C, the percentages are reduced to 0.22, 0.04, 0.23, and 0.15 %, respectively, from the original levels of 10.74, 8.58, 3.81, and 20.59 %. It can be speculated that fly ash is decomposed and volatilized in the form of chlorides and sulfides. In addition, the decomposition of Cl, K_2O, and Na_2O occurs in the whole melting process, while the SO_3 decomposition and volatilization occurs primarily between 1150 and 1260 °C.

(2) Phase constitution

According to the phase analysis of original fly ash and slag, the original fly ash contains various phases, mainly rock salt (NaCl), potassium (KCl), and $CaCl_2 \cdot Ca(OH)_2 \cdot H_2O$. High content components, such as SiO_2, Al_2O_3, and S exist in amorphous solids. When the temperature reaches 1100 °C, anhydrite ($CaSO_4$), calcium aluminum melilite ($Ca_2Al(AlSi)O_7$), and potash (KCl) become dominate phases. In particular, anhydrite is an important feature of this temperature segment, indicating that *S* is fixed in the slag in the form of $CaSO_4$. When the temperature rises to 1150 °C, the characteristic peak for anhydrite weakens, while that for calcium aluminum melilite reaches the strongest. It means that $Ca_2Al(AlSi)O_7$ stabilizes and $CaSO_4$ starts to decompose, causing increased sulfur content in the flue gas. At a temperature of 1350 °C, all phases are damaged and the slag becomes amorphous-vitreous; Cl and S have almost completely evaporated.

(3) Alkalinity change

Alkalinity (k) refers to the mass fraction ratio of basic oxides and acidic oxides in fly ash, generally expressed as $k = (CaO+Fe_2O_3+MgO+K_2O+Na_2O)/(SiO_2+Al_2O_3)$.

Alkalinity is a comprehensive reflection of fly ash composition. It tends to decrease as temperature rises in the heating process, but after reaching the melting temperature, alkalinity no longer changes with temperature and stabilizes around 0.95. In the heating process, both acidic oxides and alkaline oxides increase as there is large-scale decomposition of CaO, MgO, K_2O, and Na_2O and volatilization of S and Cl. Nevertheless, the alkalinity declines as a result due to little decomposition and volatilization of acidic oxides. When the temperature reaches the melting point, the contents of acidic and alkaline oxides are close in the slag (i.e., k reads about 1) and when above the melting point, alkalinity changes little as volatilization weakens.

(4) Volatilization rate and volume reduction rate

The volatilization rate and volume reduction rate reflect the intensity of reactions in fly ash melting and solidification process, but also as important indicators, measure the melting and solidification results. In the temperature range of 800–1150 °C, there is a little change of the volatilization rate. When the temperature rises from 1150 to 1260 °C, the volatilization rate rockets from 13.5 to 33.8 %, an increase of 23.3 %, and the change slows down at temperature above 1260 °C. The change of volume reduction rate is observed in the temperature range of 1150–1260 °C, up from 13 to 85 %. When the temperature ranges between 1260 and 1350 °C, neither the volume nor the quality of the slag changes much.

At temperature below 1150 °C, the volatilization is mainly caused by low melting point chlorides (e.g., $CaCl_2$, KCl, and NaCl) and low boiling point heavy metals (e.g. Pb and Cd). In the temperature range of 1150–1260 °C, there are violent reactions between fly ash components. Chlorides are almost completely decomposed and sulfur is involved in reactions ($CaSO_4$) and volatilized into the flue gas, resulting in increased amounts of volatile ash. It can be speculated that, the higher Cl and S contents, the higher volatilization rate of fly ash. The speculation is proven by the change of volume reduction rate. Fly ash does not melt until the temperature reaches 1260 °C. It means that, the volatilization and volume reduction of salts and metals occurs in the temperature range of 1150–1260 °C.

(5) Metal fixation rate

In the melting process, the majority of heavy metals are fixed in the slag, greatly weakening the leaching characteristics to meet environmental requirements. The rest heavy metals are volatilized in gaseous form, increasing the load on the flue gas treatment and causing potential secondary pollution. For heavy metals, the fixation rate can be expressed as follows:

$$k = \frac{m_2 c_2}{m_1 c_1} \times 100\ \%$$

where k represents the fixation rate (%); m_1 stands for fly ash quality (g), and m_2 slag quality (g); c_1 and c_2 indicates the concentrations of heavy metals before and after melting fly ash (ng/g), respectively.

According to the analysis results, at 1350 °C, the fixation rate is high for Cr and Zn, reaching 94.2 and 81.7 %, respectively, but low for Cu, Pb, and Cd, registering 31.4, 14.5, and 24.6 %, respectively. It implies that melting greatly changes the migration characteristics of heavy metals. Heavy metals with high boiling point, such as Cr and Zn, are largely fixed in slag, while volatile metals, such as Cu, Pb, and Cd, emit in the flue gas. The results echo with Jakob's findings. Considering the high chlorine content and low melting point and boiling point of metal chlorides, it can be drawn that Cu, Pb, and Cd convert into volatile chlorides and Cr and Zn form stable high silicon aluminum oxides at high temperatures, thereby suppressing decomposition and volatilization.

(6) Leaching of heavy metals

Leaching is one of the main reasons that fly ash is categorized into hazardous waste, and the toxicity of leached slag is an important indicator to measure fly ash treatment. In the analysis using horizontal oscillation and toxicity characteristic leaching procedure (TCLP), there are low amounts of Cr, Cd, Pb, Cu, and Zn leached at 1350 °C, indicating that the majority are fixed in the vitreous, which is consistent with the findings of Donald et al. In addition, the leached amount varies between metals. Among the five metals, the amounts of Zn leached differ sharply using the two approaches because of its instability under acidic conditions like the leaching solution in TCLP approach (pH = 4.3). It should be noted that the leaching concentration of heavy metals in the fly ash are below the limits for hazardous waste, so the slag is considered nonhazardous waste and potentially useful resources.

4.2.1.3 Fly Ash Melting Agents

(1) Mechanism of action

(a) B_2O_3
B_2O_3 constitutes the glass structure and can reduce the viscosity of molten vitreous, thereby reducing the ash flow temperature and melting temperature. At the same time, B_2O_3 is conducive to the diffusion of Ca^{2+} in the liquid phase to the Fe_2O_3 surface and facilitates the formation of calcium ferrite. In the presence of calcium ferrite that increases eutectic products with low melting point, the ash melting temperature and flow temperature are reduced. Studies have shown that, when added together, B_2O_3 and MgO can complement each other in vitreous performance and improve physical properties of vitreous by controlling the glass phase content and strengthening the glass structure.

(b) MgO
MgO is an intermediate of the vitreous network that builds the structure of vitreous silica. Practice has proved that the appropriate application of MgO is favorable for vitreous formation, strength, and liquidity. Meanwhile, MgO can reduce the formation of calcium ferrite and increases glassiness, which inevitably weakens the weathering intensity. When the MgO content exceeds a certain limit, MgO plays the role of modifier which undermines the formation and strength of vitreous.

(c) TiO_2
TiO_2 is an intermediate of the vitreous network. At high temperatures, Ti^{4+} exists mainly in [TiO6] 8^- in one phase and does not participate in network formation. TiO_2 changes the vitreous structure by significantly changing the interfacial energy of the concentration phase. It weakens the viscosity of molten bath and increases the diffusion rate of migrant ions, lowering the ash melting temperature and flow temperature. In addition, TiO_2 is an effective nucleating agent that promotes the formation of vitreous.

(d) WO_3
WO_3 is a surfactant that facilitates melting, but has little effect on the vitreous performance. WO_3 especially tungsten tailings, can be used to flux glass. It has been noted that WO_3 is an active agent promoting silicate reaction. As far as tungsten tailings as concerned, WO_3 and Fe_2O_3 facilitate quartz crystal transition as mineralizing agents and CaO, Na_2O, and K_2O can reduce glass viscosity at high temperatures, making sands easy to dissolute and diffuse and making it easy to remove bubbles. Tungsten tailings are waste materials of production and the utilization is considered environment friendly recycling that conforms to green production and is completely possible in production.

(e) CaF_2
CaF_2 is generally used as a nucleating agent and it can impede the formation of crystal nucleus and growth of mineral phase structure. In fly ash melting process, F^- is integrated into SiO_3F or AlO_3F and when it enters the glass

lattice, $[SiO_4]^{2-}$, a basic ingredient of mineral phase structure is replaced by SiO_3F. The reduction of $[SiO_4]^{2-}$ represents a lower possibility of mineral phase formation. However, CaF_2 and B_2O_3 can make a single crystal structure and simulate the formation and growth of crystal nucleus, though some researches hold they have no such influence.

(2) Effects

The effects of melting agents are verified in a secondary combustion chamber of an incineration plant in Fuzhou and in a baghouse of an incineration plant in Shanghai, covering melting temperature, volatilization rate, fixation rate of heavy metals, and residence time.

(a) Melting temperature

Melting agents can significantly reduce the flow temperature of fly ash. In the case of Shanghai, a 10 % addition of SiO_2 can lower the flow temperature by 80 °C, while a 10 % addition of complex additives can reduce the temperature by up to 160–1100 °C from the original level of 1260 °C. In the case of Fuzhou, by adding 10 % complex additives, the flow temperature is reduced by 130 °C from 1250 to 1120 °C. It is visible that the effects of complex additives are obvious (Figs. 4.5 and 4.6).

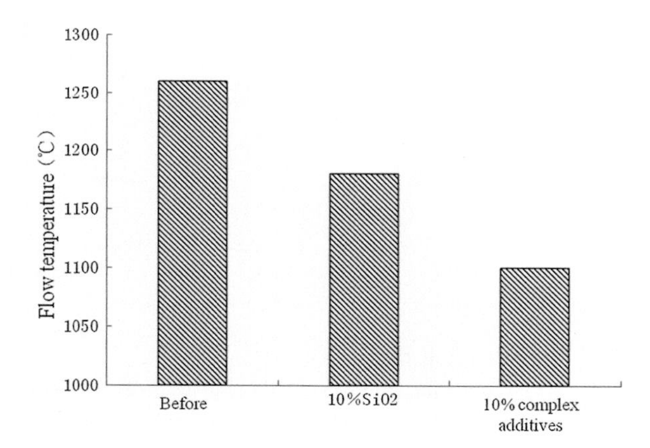

Fig. 4.5 Effect of melting agents on the flow temperature (Shanghai)

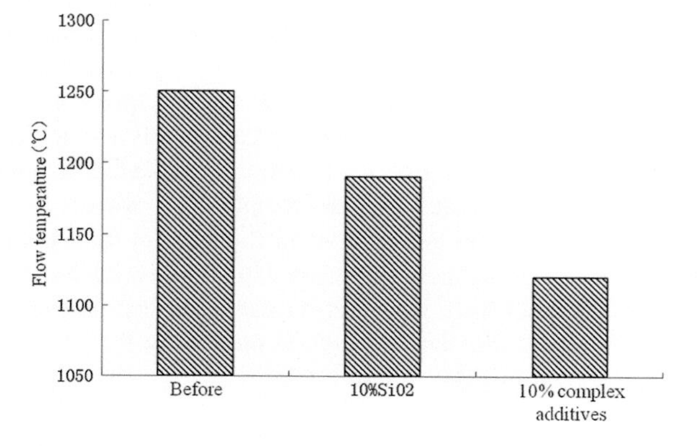

Fig. 4.6 Effect of melting agents on the flow temperature (Fuzhou)

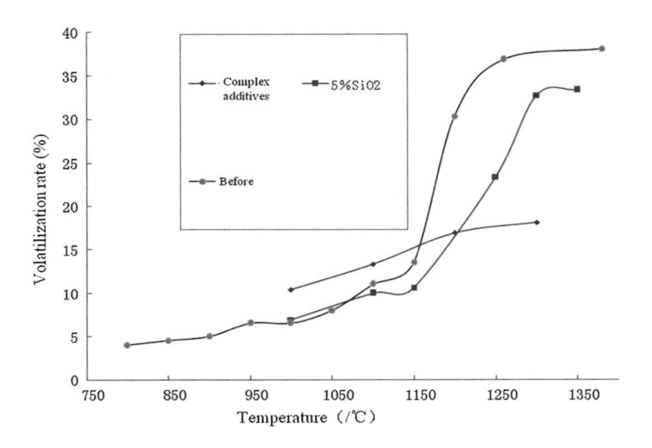

Fig. 4.7 Effect of melting agents on the volatilization rate (Shanghai)

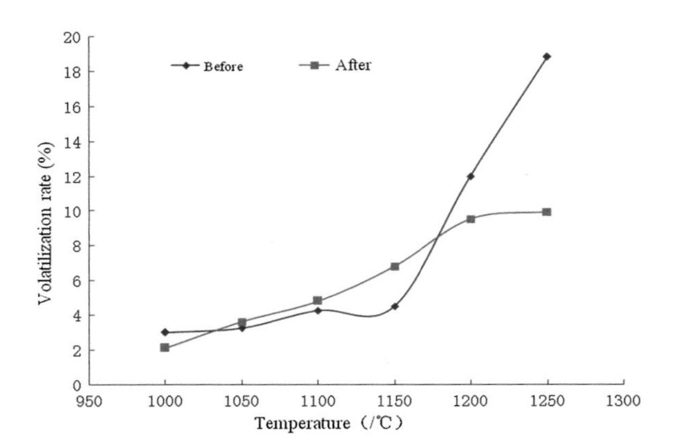

Fig. 4.8 Effect of melting agents on the volatilization rate (Fuzhou)

(b) Volatilization rate

In the melting process, the volatilization rate of fly ash shows similar trends regardless of SiO_2. At temperature lower than 1150 °C or higher than 1250 °C, the volatilization rate changes little, and when the temperature ranges between 1150 and 1250 °C, the volatilization rate increases rapidly. Analysis found that in the presence of SiO_2, the melting temperature is about 1220 °C, proving that volatilization occurs at certain temperature before melting.

However, complex additives can greatly change the volatilization rate. In the presence of complex additives, fly ash volatilization maintains stable with no sharp increase and at temperature lower than the melting point, the volatilization rate is close to the level of fly ash without complex additives. It means that, melting agents can suppress volatilization at the melting temperature but have no effect on the volatilization rate before reaching the melting temperature. The results are proved by fly ash data in the case of Fuzhou, as shown in Figs. 4.7 and 4.8.

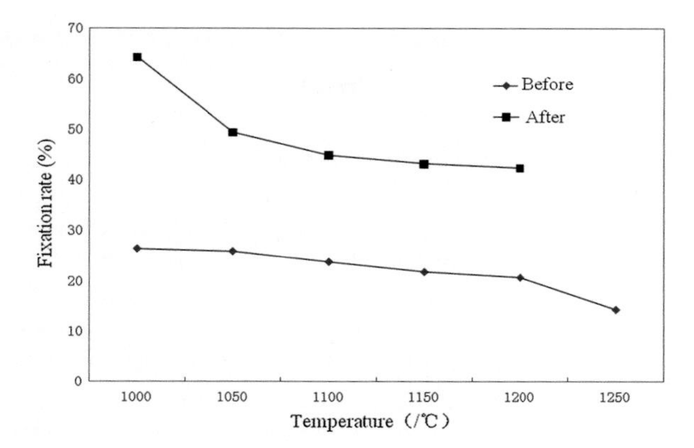

Fig. 4.9 Effect of melting agents on Cu fixation rate (Shanghai)

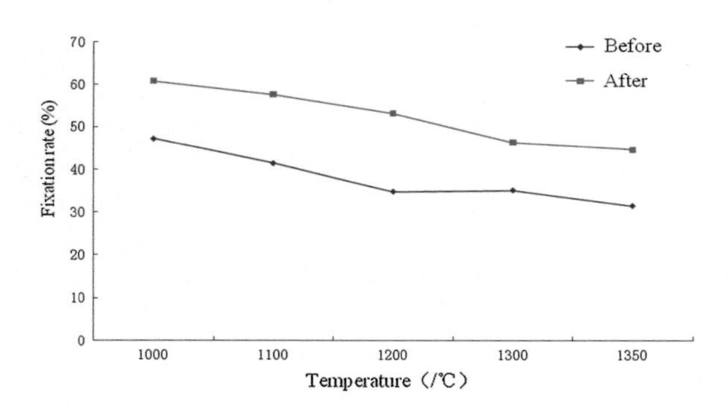

Fig. 4.10 Effect of melting agents on Cu fixation rate (Fuzhou)

(c) Fixation rate of heavy metals

Melting agents can significantly increase the fixation rate of metal Cu in the melting process. As shown in Fig. 4.9, at 1000 °C, the fixation rate is increased to 64, 38 % higher than that of scenario without additives. At a temperature of 1050–1200 °C, the fixation rate is in steady decline, but keeps 25 % higher than that of scenario without additives on average. It is substantially the same trend in the case of Shanghai, as shown in Fig. 4.10.

In the heating process, the Pb volatilization rate is large, and can reach 80 % at 1000 °C in the absence of additives. Additives can significantly improve the fixation rate. In the case of Fuzhou, the fixation rate is increased from 18.8 to 53.5 % at 1000 °C, and remains 20 % higher than that of scenario without additives at temperature 1000–1200 °C. In the case of Shanghai, there is only a slight increase of fixation rate when the temperature stays between 1100 and 1350 °C.

In short, melting agents greatly reduce the flow temperature and stimulate volatilization, enhancing the fixation of heavy metals, but the role in cutting energy consumption and treatment costs is limited.

4.2.2 Synergetic Hazardous Waste Treatment in Cement Kiln

4.2.2.1 Technological Overview

In the cement industry, waste is treated mainly in four ways. (1) Waste materials are used in clinker formation by means of calcinations as secondary raw materials or secondary fuels in the cement kiln. (2) Cement admixture is ground together with cement clinker or separately. (3) Hazardous waste is incinerated in cement rotary kiln incineration. (4) Hazardous waste, particularly radioactive waste, is fixated in cement (Xie 2010).

The process of synergetic hazardous waste treatment in cement kiln reduces the environmental impact caused by solid waste and changes waste to resources, providing energy and resources for the cement industry. Therefore, it has been unanimously approved and widely used at home and abroad. In the cement industry, success has been achieved in the treatment and utilization of a variety of waste with certain activity and similar components as cement, such as fly ash and blast furnace slag. The waste is widely used in supplementary materials and cementitious materials and becomes the important raw material for industrial production of cement. China remains in the initial stage of collaborative treatment of cement kiln, covering hazardous waste, municipal waste (household waste, sludge, etc.), and contaminated soil, as well as general industrial solid waste containing organic matter.

4.2.2.2 Synergetic Treatment

By way of synergetic treatment, waste is used in cement production process to replace the primary fuels and raw materials, to achieve synergies between cement production and hazardous waste disposal. Given the high combustion temperature (above 1600 °C), materials stay in the kiln system for 20–30 min and run under negative pressure, and flue gas residence time is more than 5 s. Under such stable conditions, pollutants can be completely burned, heavy metals effectively fixed, and a variety of toxic wastes effectively degraded. Synergetic hazardous waste treatment provides a better option for the cement industry and waste management. As far as cement industry is concerned, this process cuts the consumption of fuels and raw materials and facilitates cost-effective production. In terms of waste management, this option does not require the establishment of specialized waste incinerators or waste landfills, saving land and financial inputs. This approach optimizes waste management in ecological, social, and economic aspects, and it is a useful complement to waste disposal that produces considerable social, economic, and environmental benefits.

In China, several cement enterprises have carried out a certain scale of business continuously for the synergetic disposal of hazardous waste and municipal waste, while some introduced intermittent, small-scale business for either hazardous waste disposal or municipal waste disposal, such as experimental work on collaborative disposal of MSWI fly ash, waste pesticides, waste clay, and heavy metals. For example, Beijing Cement Plant has used dyes and paints as an alternative fuel and soil with heavy metals and organic pollutants as an alternative raw material. Guangzhou Zhujiang Cement Plant has used waste leather as an alternative raw material and plans to use MSWI fly ash as an alternative raw material. Chongqing Lafarge Cement Plant also launched a collaborative sludge disposal program. Therefore, as the advancement in theory and practice and the improvement of relevant laws and regulations, the synergetic disposal of hazardous waste in cement kiln will stand out by virtue of the economic and environmental advantages and play an increasingly important role in resourcization and harmless disposal of hazardous waste.

4.2.2.3 Development Trend

The synergetic treatment process exhibits three characteristics. (1) The resource properties of waste can be manifested to the largest degree. Not only the energy of waste can be fully utilized (alternative energy), but also waste itself can replace a portion of raw material (alternative raw materials). (2) Pollutant generation and emission can be inhibited and reduced. Given high temperature and long residence time, the organic waste can be completely decomposed. Under alkaline conditions, such acid gases as SO_2 and HCl and dioxins are effectively suppressed. Moreover, the incineration of organic waste does not influence the NO_x production in cement kilns, and causes smaller NO_x emissions compared with dedicated incineration. (3) The scope of application is wide. The process can be used to dispose organic waste and inorganic waste, free from the influence of waste characteristics due to large heat capacity. Heavy metals can be fixed to some extent, and no residues are left, avoiding follow-up treatment. However, the limitations should also be noted. The technical application requires suitable cement production facilities in the vicinity of waste. The process is not applicable to the disposal of waste with high content of heavy metals because heavy metals cannot be decomposed or removed, or completely solidified. The disposal of waste with chlorine and sulfur content is also undermined, due to the restrictions in cement production process and cement quality.

References

Bernardo E. Micro- and macro-cellular sintered glass-ceramics from wastes. J Eur Ceram Soc. 2007;27(6):2415–22.

Comstock SE, Boyer TH, Graf KC, Townsend TG. Effect of landfill characteristics on leachate organic matter properties and coagulation treatability. Chemosphere. 2010;81(7):976–83.

He PJ, Zhang H, Wang ZD, Zhang CG. Pollution characteristics of air pollution control residues from municipal solid waste incineration plant. J Tongji Univ. 2003;31(8):972–6.
Hu WT, Zhang J. Review of hazardous waste treatment and disposal. J Anhui Agric Sci. 2014a;42(34):12386–8.
Hu W, Zhang J. Review of the status quo of hazardous waste treatment and disposal. J Anhui Agric Sci. 2014b;42(34):12386–8.
Lombardi F, Mangialardi T, Piga L, Sirini P. Mechanical and leaching properties of cement solidified hospital solid waste incinerator fly ash. Waste Manag. 1998;18(2):99–106.
Mangialardi T, Paolini A, Sirini P. Optimization of the solidification/stabilization process of MSW fly ash in cementitious matrices. J Hazard Mater. 1999;70(1):53–70.
Muller M, Milori DMBP, Déléris S, Steyer J-P, Dudal Y. Solid-phase fluorescence spectroscopy to characterize organic wastes. Waste Manag. 2011;31(9):1916–23.
Quina MJ, Bordado JC, Quinta-Ferreira RM. Treatment and use of air pollution control residues from MSW incineration: an overview. Waste Manag. 2008;21:313–23.
Rad BA, Alizadeh P. Pressureless sintering and mechanical properties of SiO_2-$A1_2O_3$-MgO-K_2O-TiO_2-F (CaO-Na_2O) machinable glass-ceramics. Ceram Int. 2009;35(7):2775–80.
Ubbnaco P, Calabrese D. Solidification and stabilization of cement paste containing fly ash from municipal solid waste. Thermochim Acta. 1998;321(1–2):143–50.
Xie Y. Study on the technology of co-processing MSWI ash in cement industry and solidifying mechanism of heavy metal. Guangzhou: South China University of Technology; 2010. p. 1–2.
Zeng CM, Li X, Wu JJ, Huang HY. Research on MSWI fly ash disposal. J Dongguan Univ Technol. 2012;19(3):73–6.

Chapter 5
Groundwater Pollution and Its Risk in Solid Waste Disposal Site

Abstract Landfill, the eventual way of disposing solid wastes, inevitably comes with a certain degree of pollution to surrounding groundwater due to construction, broken impervious layer, and uneven geological subsidence. The survey and risk assessment of groundwater pollutions are the important technical means for knowing the current status of landfill site pollution, forecasting the pollution trend, and assessing its risk. This chapter elaborates the general procedures for survey of groundwater pollution on various landfill sites, including the collection of basic data, setting out of monitoring points, selection of monitoring items, and data organization method, according to China's current mode of construction and operation of landfill sites, in combination with the research progress in groundwater survey both domestic and foreign. On this basis, the typical model—3 MRA risk assessment model—is optimized.

Keywords Household waste landfill site · Hazardous waste landfill site · Groundwater · Pollution survey · Risk assessment

5.1 Groundwater Pollution in Solid Waste Disposal Site

5.1.1 Household Waste Landfill Site

At present, nationwide, domestic household waste landfill sites are mainly concentrated in northeast and north China as well as areas in Jiangsu and Zhejiang Provinces and along the middle and lower reaches of the Yangtze River, which are all densely populated. The sites are distributed generally more in southeast China than in northwest China. The landfill sites were largely built 5–10 years ago. Of them, 20 % became available after 2010. This shows rapidly growing landfill sites for nearly 10 years, during which industrial production expands significantly and the living standard of people improves. In the survey, most of the landfill sites are rated level-II, basically acceptable in terms of harmless disposal. There are,

B. Xi et al., *Optimization of Solid Waste Conversion Process and Risk Control of Groundwater Pollution*, SpringerBriefs in Environmental Science,
DOI 10.1007/978-3-662-49462-2_5

however, about 1/3 level-IV landfill sites, where wastes are disposed simply and greatly harmful to environment (Figs. 5.1 and 5.2).

From the perspective of existing monitoring wells for waste landfill sites, nearly 70 % of the sites have no monitoring well, and only 10 % meet the requirements for the number of monitoring wells as specified in GB16889-2008 *Standard for Pollution Control on the Landfill Site of Municipal Solid Waste*. This indicates that the majority of existing normal landfill sites virtually failed to satisfy the requirements for continuous monitoring of groundwater pollution. Moreover, the monitoring wells are set out without considering the groundwater flow direction and hydrogeological characteristics, and in general, the monitoring index only includes total hardness, ammoniacal nitrogen, total coliform group, and COD, while heavy metals and organic pollutants are almost left non-monitored. Among the landfill sites

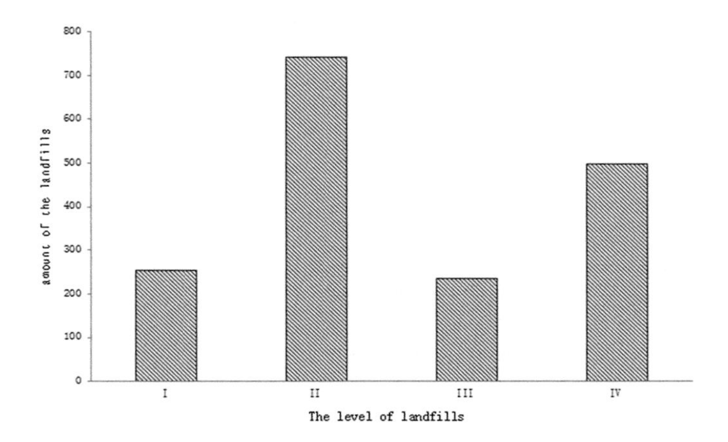

Fig. 5.1 Level of household waste landfill sites in China

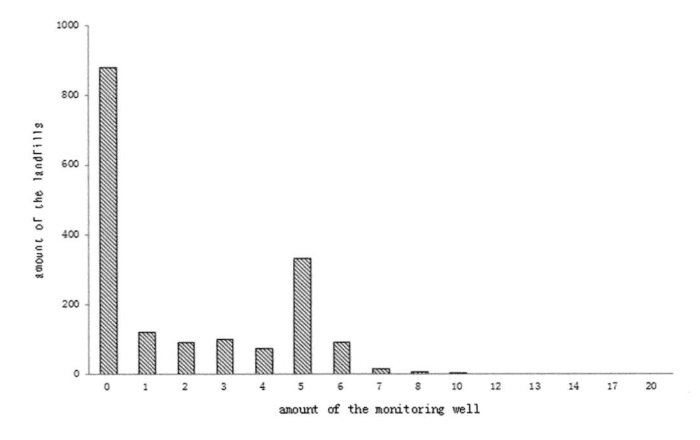

Fig. 5.2 Number of monitoring wells for household waste landfill sites in China

surveyed, 90 % had no hydrogeological data, with the direction of groundwater flow and the depth of monitoring wells unknown. The waste on many landfill sites is buried deeper than groundwater depth—a sign of a great deal of wastes being present in shallow aquifer and considerably affecting groundwater (Figs. 5.3, 5.4 and 5.5).

The groundwater in waste landfill sites is contaminated mainly by the hazardous and noxious substances in waste entering the groundwater in the form of leachate generated through anaerobic fermentation. The inspection reveals that the landfill leachate contains huge amount of poorly biodegradable noxious pollutants, which can concentrate locally and create unsafe cumulative effect. There are data to suggest that the main contamination indexes for the groundwater in China's household waste landfill sites are COD, ammoniacal nitrogen, total coliform group, heavy metal, total hardness, and oily (Khanbilvardi et al. 1995).

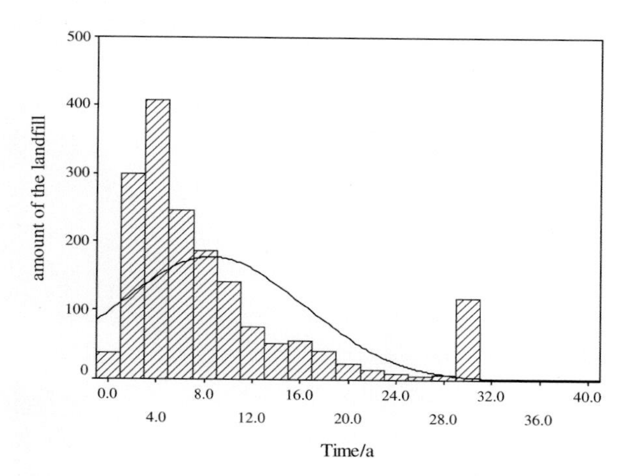

Fig. 5.3 Time of waste landfill sites

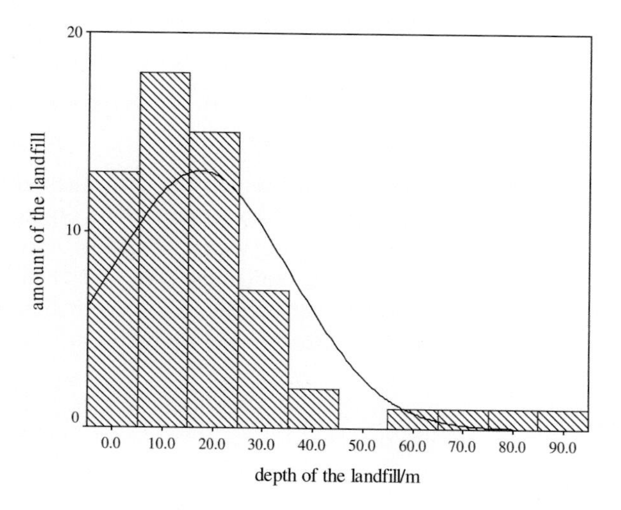

Fig. 5.4 Depth of waste landfill sites

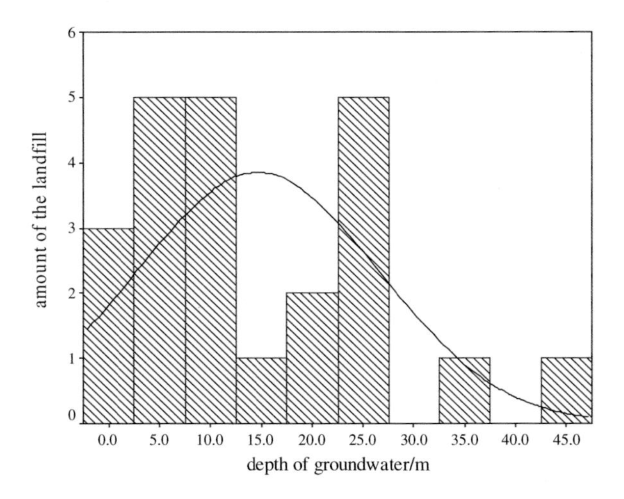

Fig. 5.5 Groundwater depth on waste landfill sites

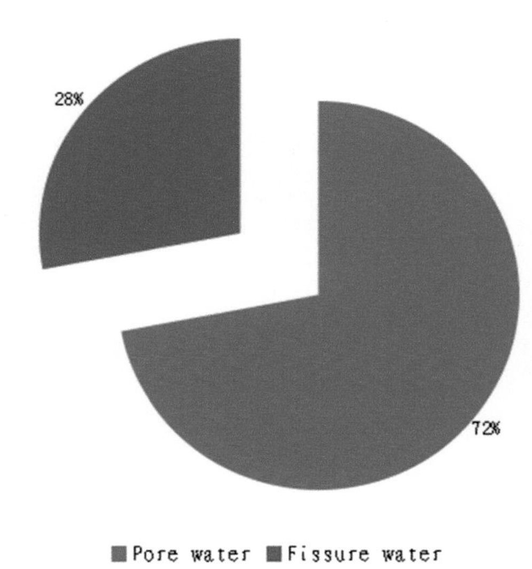

Fig. 5.6 Types of groundwater in waste landfill site

Despite the application of sanitary landfill method on nearly all waste landfill sites, the leak of landfill leachate remains. The groundwater in most of the waste landfill sites in China is pore water (Fig. 5.6), the lithology of whose aquifer features loose sediments prominently (Fig. 5.7), which are prone to pollute the groundwater. In addition, the presence of the great number of simple landfill sites is also a major reason for groundwater pollution in landfill sites.

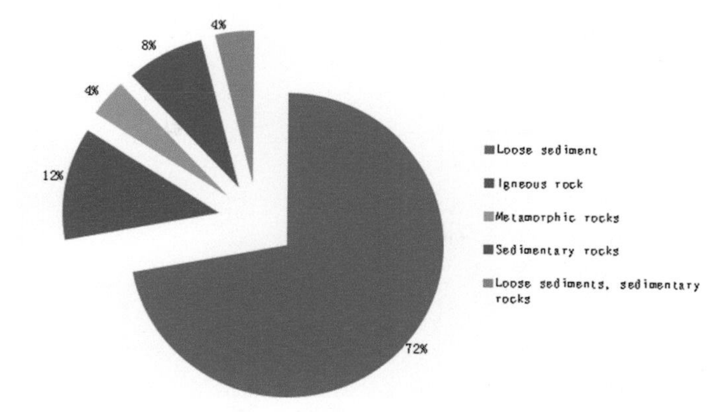

Fig. 5.7 Lithology of the aquifer in waste landfill site

5.1.2 *Hazardous Waste Landfill Site*

(1) General distribution

Chinese provinces (autonomous regions and municipalities) have totally launched 44 hazardous waste disposal and landfill center projects, including those under construction and to be constructed. At present, there are 11 hazardous waste disposal and landfill centers completed, respectively, in Shenzhen, Huizhou, Fuzhou, Hangzhou, Taizhou, Qingdao, Shenyang, Lanzhou, Chongqing, Tianjin, and Shanghai. All other Chinese provinces (autonomous regions and municipalities) except Tibet are planning to build a minimum of one such center. Of them, Inner Mongolia, Liaoning, Shandong, Shanxi, Hubei, Anhui, Hunan, Chongqing, Xinjiang, Yunnan, Zhejiang, Guangdong, and Henan have a plan to construct two or three.

(2) Level of both disposal and construction

Upon completion and operation of the 44 disposal and landfill centers, the capacity of hazardous waste disposal in China will reach to 1.9122 million t/a, of which 984,300 t/a hazardous wastes will be disposed by landfill. According to the disposal capacity and landfill capacity of Chinese hazardous waste disposal and landfill centers (see Figs. 5.8 and 5.9), 80–90 % of these centers have the capacity of 20,000–50,000 t/a for disposal and 10,000–30,000 t/a for landfill, with the amount of the wastes buried about 52.5 % higher than that disposed, indicating that secure landfill remains the main method for hazardous waste disposal and landfill centers in China. The 44 disposal and landfill centers totally cover an area of 5.802 million m^2 as planned, most of them with an area of about 100,000 m^2 and service life of generally 10–20 years.

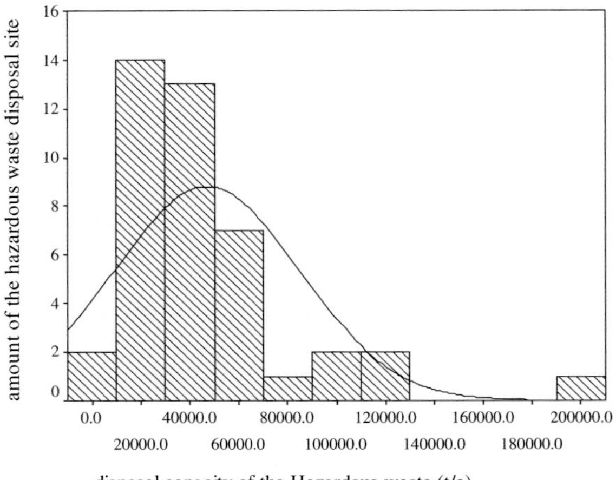

Fig. 5.8 Disposal capacity of hazardous waste in China

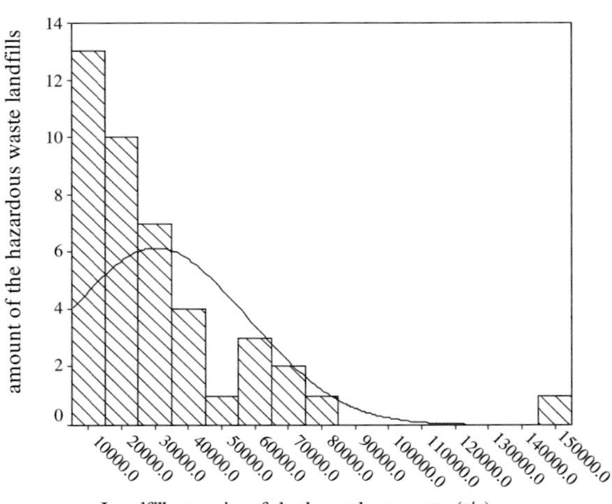

Fig. 5.9 Landfill capacity of hazardous waste in China

(3) Hydrogeological condition

The site topography and groundwater depth are two major factors for representing the environmental sensitivity to hazardous waste sites. Topography is connected closely with geological conditions, seepage supply, and hydraulic conductivity, while groundwater depth directly determines the travel distance of pollutants into aquifer. Proper selection of topography and groundwater depth is conducive to reduce the groundwater pollution in hazardous waste disposal and landfill center and achieving better disposal of hazardous wastes.

Hazardous waste disposal and landfill centers should be sited following the guideline of “suitability and feasibility” and in areas where geological structure is

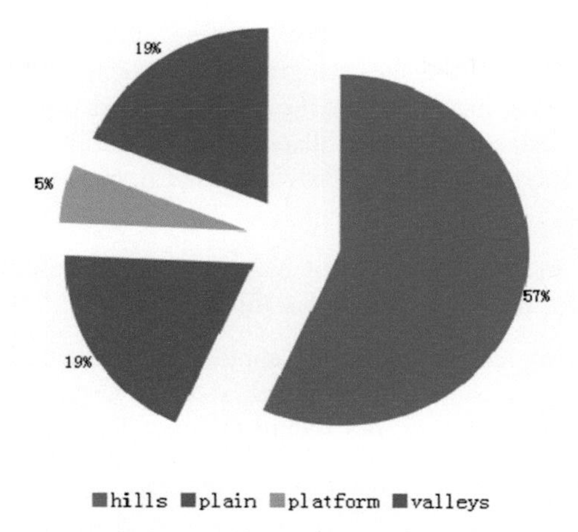

Fig. 5.10 Topographic features for hazardous waste disposal and landfill sites in China

stable, the disturbance by human activities is minor, and major accidents are unlikely to occur due to natural and human factors. Figure 5.10 shows the topographic features where the 44 hazardous waste disposal and landfill center projects in China locate. 56.8 % of the centers are built on hills, for which hill characterizes China's landscape, and above all, hilly areas are scattered with natural "landfill pits," which require less quantities of earthwork and stonework and less costs. The centers on plain and valley comprise 18.9 %, while those on platform comprise only 5.4 %.

As an important index for groundwater vulnerability assessment, groundwater depth entails considerable attention during project construction. Where the groundwater is shallow, or the construction site is located at a main drinking water source or downstream of a centralized water supply well, the site should be relocated or require better anti-seepage design to prevent landfill leachate from entering groundwater and causing groundwater pollution.

Figure 5.11 shows the groundwater depth data collected currently in the hazardous waste disposal and landfill sites. By and large, the groundwater in the

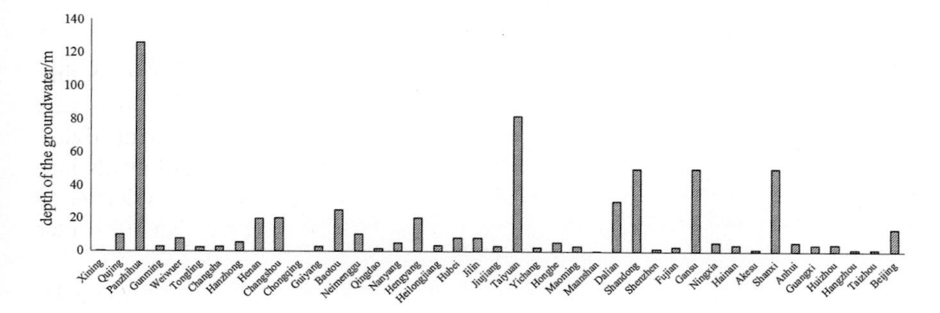

Fig. 5.11 Groundwater depth in the hazardous waste disposal and landfill sites in China

hazardous waste disposal and landfill sites in China varies widely in depth, ranging from 0.5 to 126 m. 75 % of the sites see groundwater depth of less than or equal to 10 m, and 15 of them are located where the groundwater depth is less than 3 m. The groundwater in the hazardous waste disposal and landfill sites in Panzhihua of Sichuan Province, Taiyuan of Shanxi Province, Gansu, Ningxia, and Shanxi are not less than 30 m.

(4) Classification and characteristics of hazardous waste landfill site
(i) Classification of hazardous waste landfill site

The secure landfill of hazardous wastes applies to the hazardous wastes whose components and energy cannot be recycled, such as large quantity of industrial wastes, which must be disposed in a secure manner. The requirements for secure hazardous waste landfill sites as specified in GB 18598-2001 *Standard for Pollution Control on the Security Landfill Site for HazardousWastes* are given below:

Security landfill sites must have required impermeable layer to prevent secondary pollution. If the natural material lining has saturated permeability coefficient of less than 1.0×10^{-7} cm/s, and a thickness of more than 5 m, the natural clay lining can be directly used as impermeable layer; if its saturated permeability coefficient is 1.0×10^{-7}cm/s–1.0×10^{-6} cm/s, composite lining can be used as impermeable layer, with high-density polyethylene not less than 1.5 mm thick; if its saturated permeability coefficient is 1.0×10^{-6} cm/s, it is necessary to use double synthetic linings (high-density polyethylene) for impermeable layer, with the upper and lower layers more than 2.0 and 1.0 mm thick, respectively.

Currently, hazardous waste landfill sites are mainly classified into three types, rigid, flexible, and rigid–flexible combination, both at home and abroad, according to the impervious structure.

① Flexible structure

Flexible structure is generally used where the landfill site basically meets the geological requirements for its siting. The anti-seepage system of the flexible structures must be provided with double synthetic linings. The flexible structure includes, from the bottom top, the base layer, groundwater drainage layer, compacted clay lining, high-density polyethylene film, above-film protective layer, auxiliary leachate drainage layer, high-density polyethylene film, above-film protective layer, main leachate drainage layer, geotextile, and hazardous wastes. The upper high-density polyethylene film should have a minimum thickness of 2.0 mm, while the lower one should be at least 1.0 mm thick.

② Rigid–flexible combination

The landfill site which partially meets the geological conditions for siting can combine reinforced concrete shell and flexible synthetic lining to create a rigid structure for the purpose of impermeability. The structure comprises, from the bottom up, reinforced concrete floor, groundwater drainage layer, below-film composite bentonite protective layer, high-density polyethylene impermeable film, geotextile, pebble layer, and hazardous wastes. For the side walls, the anti-seepage system is composed of reinforced concrete wall, geotextile, high-density polyethylene impermeable film, geotextile, and hazardous wastes from the outside in. The synthetic lining should be constructed of materials with desirable chemical compatibility, durability, and heat resistance as well as high strength, low permeability, low maintenance, and no secondary pollution. If high-density polyethylene film is used, its permeability coefficient must be less than or equal to 1.0×10^{-12} cm/s. The high-density polyethylene film provided at the bottom and sides of rigid landfill site should have a thickness of 2.0 mm or above.

③ Rigid structure

Rigid structure is required in areas where the landfill site is basically not in compliance with the geological conditions for siting, or the buried hazardous wastes may leave environment likely exposed to pollution. All impermeable layers of such structure should be constructed of reinforced concrete. Those for the side walls and the bottom of the box should be designed as anti-seepage structure and proved acceptable for crack widths, and should have permeability coefficient of 1.0×10^{-6} cm/s or below. This is a fairly rare occurrence, and can be addressed through risk reduction by proper location of hazardous wastes and pretreatment of the hazardous wastes that present high environmental pollution risk.

(ii) Characteristics of hazardous waste landfill site

① Small size. The secure landfill is significantly smaller than the sanitary landfill site for household waste, largely due to the amount of industrial hazardous waste generated considerably less than that of household waste.

② High impermeability performance. Compared with household waste landfill site, the secure landfill should have higher impermeability performance, which is generally achieved by applying double impermeable layer system, with the HDPE impermeable film more than 2.0 mm thick. The sanitary landfill site for household waste normally requires only single impermeable layer, whose HDPE impermeable film needs to more than 1.5 mm thick simply.

③ Strict management of landfill operation. The secure landfill site should perform landfill operations in sections, and adequately keep hazardous wastes away from rain so as to avoid generation of excessive harmful leachate.

④ At present, for impermeability of secure landfill sites both domestic and foreign, flexible structure is applied more often, for it is inexpensive and technically mature, and enables flexible technological combination and high adaptability of the site to topography and hydrologic conditions.

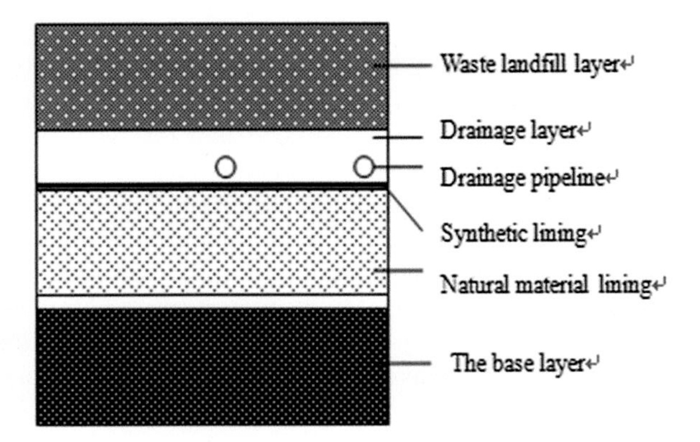

Fig. 5.12 Impermeable structure with single-layer lining

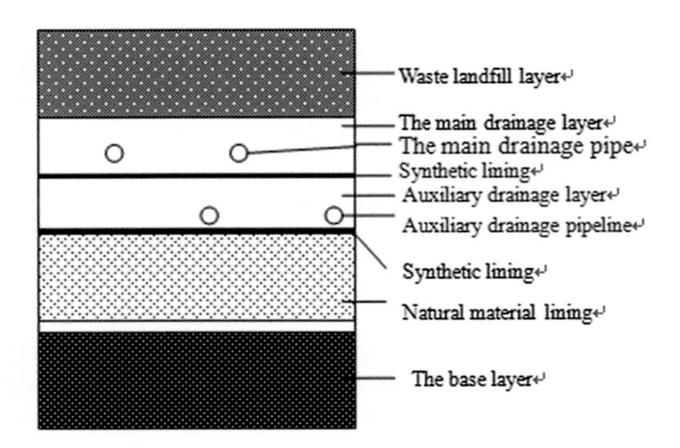

Fig. 5.13 Impermeable structure with double-layer lining

The synthetic lining for landfill sites should be constructed of materials with desirable chemical compatibility, durability, and heat resistance as well as high strength, low permeability, low maintenance, and no secondary pollution. If high-density polyethylene film is used, its permeability coefficient must be less than or equal to 1.0×10^{-12} cm/s.

The flexible structure appears mainly in two impermeable forms, single-layer lining and double-layer lining, and their section configuration is shown in Figs. 5.12 and 5.13, respectively.

The issue of whether the single-layer lining or the double-layer lining is preferable for impermeability has always been a focus of debate for experts and scholars domestic and foreign. The single-layer lining structure is easy to build and requires low investment, but needs to meet rigorous site conditions concerning engineering geology and hydrogeolology. For instance, the high groundwater level on the site should be more than 2 m away from the impermeable layer, and the clay layer below the impermeable layer should have a thickness of not less than 1 m and permeability coefficient of less than 10^{-7} cm/s. The impermeable structure featuring double-layer lining involves complicated construction, high costs, and remarkable performance in pollution prevention. The secure landfill sites in China mostly feature anti-seepage system with double-layer lining. Different landfill areas contain different hazardous wastes, with a significantly different leachate component phase generated. Strictly speaking, the anti-seepage system is designed basically to prevent leachate from contaminating soil and groundwater. Therefore, its structural design is also associated with the division of landfill areas, as the leachate with different components requires different anti-seepage structures and materials.

Generality and individuality is applicable to any project. The use of single-layer lining or double-layer lining for impermeability is subject to the specific conditions, mainly in terms of the amount of leachate generated and the composition and concentration of the pollutant contained. The double-layer lining structure is not compulsory, and when the conditions permit, the anti-seepage system with single-layer lining is also optional. The selection of the structure and materials for the lining depends on the performance of the material in preventing seepage and pollution as well as economic factors, that is, technically practical and feasible and economically rational.

(5) Pollution of groundwater in hazardous waste landfill site

The hazardous waste landfill sites completed prior to "11th Five-Year Plan" period have brought environmental pollution for many reasons. After the "11th Five-Year Plan" period began, we surveyed eight of the completed hazardous waste landfill sites subjected to the *National Hazardous Waste and Medical Waste Disposal Facility Planning*. Of the eight sites, only two sites met the Class-III standard on all monitoring indexes as specified in GB/T14848-93 Quality Standard for Ground Water, while the other six hazardous waste landfill sites saw substandard groundwater to varying degrees, primarily with respect to pH, followed by ammoniacal nitrogen. The turbidity, escherichia coli, lead, hexavalent chromium, arsenic, nickel, and manganese were also unacceptable for some sites. The hexavalent chromium in the groundwater in one hazardous waste landfill site was substandard. The survey data show that of the eight wells, five were proved unacceptable and caused adverse impact on the groundwater.

The following may be the reasons for the pollution:

(i) Broken anti-seepage system in the landfill sites.
(ii) Unreliable monitoring data, and the rainwater or other polluted water collected due to improper method used for water sampling in the monitoring wells. The results cannot represent local groundwater quality.
(iii) As the background value for local groundwater was unavailable, the local groundwater may be polluted by other pollutant, and showed high background value.

(iv) Upstream pollution.
(v) The background value of groundwater was high due to the pollution from surface runoff or agricultural irrigation water.

5.2 Survey of the Groundwater Pollution in Solid Waste Disposal Site

5.2.1 Survey Content and Procedures

In recent years, with the heightening of public awareness of environmental protection, the groundwater pollution by solid wastes has been drawing wide attention. It is imperative to carry out groundwater survey near landfill sites, in order to provide essential data for identification, characterization, and management of pollution risk in the future. According to the current construction and operation mode of landfill sites in China, the groundwater survey mainly consists of early survey, data collection, survey of current status, and long-term groundwater monitoring.

5.2.1.1 Early Survey

The early survey is necessary for preparing survey of groundwater in landfill sites, and enables smooth and orderly completion of groundwater survey and monitoring activities. The field survey of landfill sites covers the basic information of the site, hydrological and meteorological characteristics, geological characteristics, and environmental characteristics, as shown in Tables 5.1 and 5.2.

Tables 5.1 and 5.2 shall be filled in strictly according to the facts during site survey. In addition, pictures showing operating conditions of landfill area shall be collected, especially pictures of sensitive points with potential risk or where contamination accident is easy to occur. The data collected are mainly used for the compilation of existing situation investigation report.

5.2.1.2 Data Collection

Basic operating conditions of landfill area are known after the existing situation investigation, but it is hard to meet the requirements of subsequent simulation and prediction, hazard identification, and risk characterization. The main aim of data collection is to collect the feasibility study report, environmental impact assessment report, engineering geological investigation report, plan sketch of monitoring wells (springs), and historical monitoring data regarding the landfill area. The survey information can be supplemented and perfected by referring to the feasibility study report, environmental impact assessment report, and engineering geological investigation report when necessary.

Table 5.1 Basic information of landfill site

<table>
<tr><td colspan="3">1. Name of landfill site (seal)</td></tr>
<tr><td colspan="3">2. Code of landfill site: □□□□□□ - □□□ - □□□□ - a</td></tr>
<tr><td colspan="3">3. Geographic position:
County (district and city), region (city, state), province (autonomous region and municipality)
4. Geographical coordinates:
Central latitude° ′″ East longitude;
Central longitude) °′″North latitude;</td></tr>
<tr><td colspan="3">5. Date of formal operation: □□□□ Y □□</td></tr>
<tr><td colspan="3">6. Date of expansion and upgrading □□□□ □□</td></tr>
<tr><td>7. Site area (m^2)</td><td colspan="2">8. Area of landfill site (m^2)</td></tr>
<tr><td>9. Depth of landfill site: (m)</td><td colspan="2">10. Capacity of landfill site (ton/year)</td></tr>
<tr><td>11. Landfill capacity: (t)</td><td>12. Service life: (year)</td><td>13. Gradient of side slope:</td></tr>
<tr><td colspan="3">14. Structure of impermeable layer: □Rigid □Flexible
Bottom: □Natural clay□Single-layer synthetic lining material □Double-layer synthetic lining material
Side slope: □Natural clay□Single-layer synthetic lining material □Double-layer synthetic lining material</td></tr>
<tr><td colspan="3">15 Management</td></tr>
<tr><td colspan="3">15.1 Name of management unit:</td></tr>
<tr><td>15.2 Number of personnel: _ (person)</td><td colspan="2">15.3 Waste collection: □Stable □ Unstable</td></tr>
</table>

5.2.1.3 Simulation and Prediction

By applying the groundwater simulation software such as Visual Modflow and GMS, a conceptual model of groundwater in the landfill area and a water current model are built according to the previous survey and investigation data and the obtained hydrogeological environment information regarding the landfill area for the purpose of predicting the migration tendency and diffusion range of typical contaminants from the landfill area.

5.2.1.4 Supplementary Investigation of Current Situation

Compared with previous survey and investigation, the ongoing site survey shall place the emphasis on the supplementary collection of data necessary for simulation and predication. The information contained in preliminary geological survey report, feasibility study report, and environmental impact assessment report made at early stage of landfill site engineering design can be used as historical observation data of the site. As current hydrogeological conditions, groundwater quality, and sensitive points have changed, it is necessary to conduct a supplementary investigation on existing situation of groundwater quality to meet the requirements of Level I evaluation criterion in *Technical Guidelines to Environmental Impact Assessment—Groundwater Environment*.

Table 5.2 Hydrogeological environment information survey on landfill site

<table>
<tr><td colspan="6">1. Hydrological characteristics</td></tr>
<tr><td colspan="2">1.1 Groundwater type</td><td colspan="2">□ Phreatic water □ Confined water □ Pore water □ Fissure water □ Karstic water</td><td colspan="2">1.2 Purpose of groundwater: □ Industry □ Agriculture □ Living □ Others</td></tr>
<tr><td>1.3 Antifouling property of vadose zone</td><td colspan="5">Single-layer thickness of rock (soil) stratum: _____ (m)
Osmotic coefficient: _____ (m/d)
Whether the distribution is continuous and stable: □ Yes □ No.</td></tr>
<tr><td>1.4 Aquifer</td><td colspan="5">Burial depth:_____(m); thickness: _____(m); hydraulic gradient: ______;
annual water level amplitude: _____(m); specific yield: _____;
osmotic coefficient: _____ (m/d)</td></tr>
<tr><td colspan="6">1.5 Water characters: Temperature: ____ (℃); pH:_____; Conductivity: _______(μs/cm); DO: _____ (mg/L); redox potential: _______ (mv); turbidity: _______(NTU); odor and taste: level ______</td></tr>
<tr><td colspan="6">2. Meteorological characteristics</td></tr>
<tr><td colspan="2">2.1 Climate type: _____________</td><td colspan="2">2.2 Annual average temperature: ______℃</td><td colspan="2">2.3 Rainy months: ______</td></tr>
<tr><td colspan="3">2.4 Average annual precipitation ______(mm)</td><td colspan="3">2.5 Average annual evaporation capacity: ______(mm)</td></tr>
<tr><td colspan="6">3. Geologic characteristics</td></tr>
<tr><td colspan="6">3.1 Landform: □ plain □ mountain land □ platform □ hill □ basin □ others</td></tr>
<tr><td colspan="6">3.2 Geological phenomenon and its orientation: □ Gully ( ) □ Collapse ( ) □ Landslide ( ) □ Fault ( ) □ Karst ( ) □ Others ( )</td></tr>
<tr><td colspan="6">3.3 Lithology of site foundation: □ Unconsolidated sediment □ Sedimentary rock □ Metamorphic rock □ Igneous rock □ Tectonite</td></tr>
<tr><td colspan="6">3.4 Lithology of aquifer: □ Unconsolidated sediment □ Sedimentary rock □ Metamorphic rock □ Igneous rock □ Tectonite</td></tr>
<tr><td colspan="6">3.5 Rock (soil) stratum structure</td></tr>
<tr><td colspan="3">Name</td><td>Initiallandfill depth (m)</td><td>Thickness (m)</td><td>Osmotic coefficient K (m/d)</td></tr>
<tr><td colspan="3"></td><td></td><td></td><td></td></tr>
<tr><td colspan="6">4. Environment characteristics</td></tr>
<tr><td rowspan="3">4.1 Leachate</td><td colspan="3">Yield: ____ (t/d)</td><td colspan="2">Discharge: ____ (t/d)</td></tr>
<tr><td>Treatment mode</td><td colspan="4">□ Flocculation □ Hydrolytic acidification □ Anaerobic sludge process □ Biofilm process □ Oxidation ditch □ Activated sludge process □ Nanofiltration □ Reverse osmosis □ Deep adsorption □ Others</td></tr>
<tr><td colspan="5">Final disposal: □ Used for site landscaping □ Discharged when reaching the standard □ Disposed to the sewage treatment plant □ Others</td></tr>
<tr><td colspan="6">4.2 Information of sensitive points: a. Surface water body (1. river; 2. lake (pond); 3. reservoir; 4. sewage ditch; 5. others); b. Residential area; c. Natural conservation area; d. Cultivation area; e. Aquiculture area; f. Water source; g. Others</td></tr>
<tr><td>Type</td><td>Name</td><td colspan="2">Location</td><td>Distance (m)</td><td>Remarks</td></tr>
<tr><td></td><td></td><td colspan="2"></td><td></td><td></td></tr>
</table>

5.2.2 *Layout of Groundwater Monitoring Wells and Sampling*

5.2.2.1 Layout of Monitoring Points

(1) Layout principles

The layout of monitoring points for groundwater investigation shall be fully combined with simulation and prediction results of groundwater pollution in the landfill, so as to make clear of pollution status of the groundwater in the investigation sites and grasp spread range of groundwater pollution in the landfill. Therefore, status investigation of groundwater pollution in the landfill shall follow the following principles: (1) Set at least six groundwater monitoring wells in the landfill, including one background monitoring well and five spread of pollution monitoring wells; (2) fully consider representativeness of the monitoring wells and scientificity of their layout, make full use of the existing monitoring wells, and set additional monitoring wells if they fail to meet the quantity or quality requirements; (3) densify the exploration points in leakage-prone and pollution spread zones such as lining junction or folding point around the landfill; (4) properly increase or reduce the distance between the monitoring points and the landfill according to such factors as geotechnical properties and types of hydrogeological units, hydrogeological parameters, monitoring direction, etc.; (5) set the monitoring wells based on the quality status monitoring network and history monitoring situations of groundwater in the landfill area (or based on vulnerability assessment and zonation of groundwater in the area); and (6) if there are spring opening points for flow out of groundwater nearby the landfill, take those in upstream direction of the groundwater flow as the background monitoring points and those in the downstream direction of the flow as spread of the pollution monitoring points.

(2) Layout methods

Multiple factors shall be considered in the point layout methods for groundwater, such as landfill boundary contour, groundwater type, topographic characteristics, hydrogeological parameters, etc. The groundwater can be generally classified into pore water from plains and flat plateaus and karst and fissure water from mountains and hills, for which the point layout methods are as follows:

(i) Pore water from plains and flat plateaus

If any boundary of such landfill is vertical to direction of the groundwater flow or the minimum angle between them is less than 10°, the point layout shall be as such: one ground monitoring well shall be set, within 30–50 m in upstream direction of groundwater flow in the landfill; five spread of pollution monitoring wells shall be set, with one, respectively, within 30–50 m on both sides vertical to direction of the groundwater flow; and two monitoring wells shall be set at 30 m in downstream direction of the groundwater flow, with one, respectively, set within 30–50 m and at 50 m in the direction vertical to the flow, as shown in Fig. 5.14.

If the minimum angle between any boundary of the landfill and direction of the groundwater flow is greater than 10° but less than or equal to 45°, the point layout shall be as such: one ground monitoring well shall be set, within 30–50 m to upper peak boundary in upstream direction of groundwater flow in the landfill; 5–6 spread of pollution monitoring wells shall be set, within 30–50 m along downstream boundary of the groundwater flow, with an equal spacing of 50–80 m; and one monitoring well shall be set, at 80 m to lower peak boundary in downstream direction of groundwater flow, as shown in Fig. 5.15.

If boundary of the landfill is irregular, the point layout shall be as such: one ground monitoring well shall be set, within 30–50 m in upstream direction of groundwater flow in the landfill; six spread of pollution monitoring wells shall be set, with one, respectively, within 30–50 m on both sides vertical to direction of the groundwater flow; four monitoring wells shall be set in downstream direction of the groundwater flow in "rhombus" shape, with one within 5–10 m in downstream direction of the groundwater flow, and one within 30–50 m (for landfill whose filling year is less than 10 a) or 50–80 m (for landfill whose filling year is less than 10 a); and diagonal length of the "diamond" vertical to the flow direction shall be 50–100 m, as shown in Fig. 5.16.

(ii) Karst and fissure water from mountains and hills

One background monitoring well can be set within 30–50 m to boundary of the landfill in upstream direction of the groundwater flow; if there are spring opening

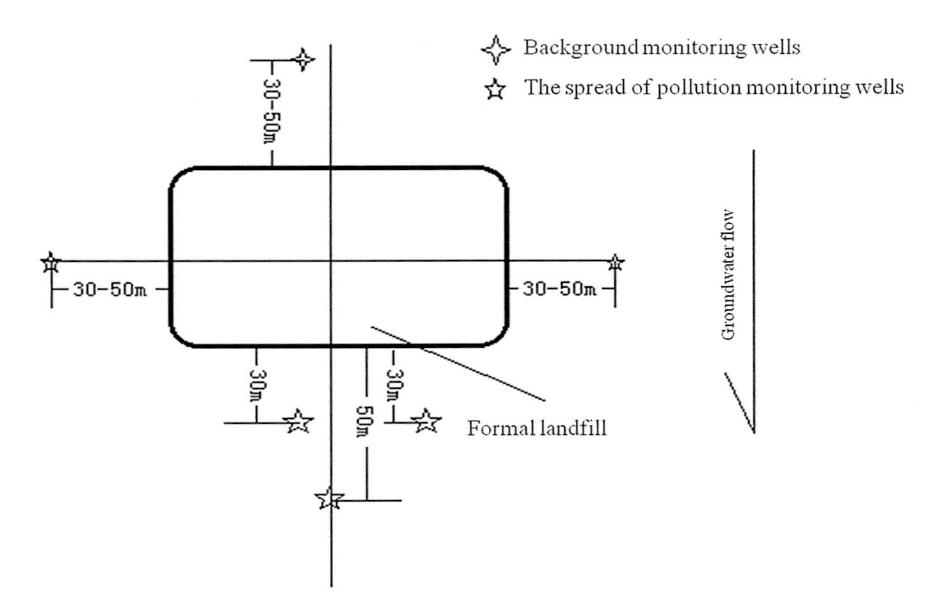

Fig. 5.14 Layout diagram of monitoring wells in the landfill

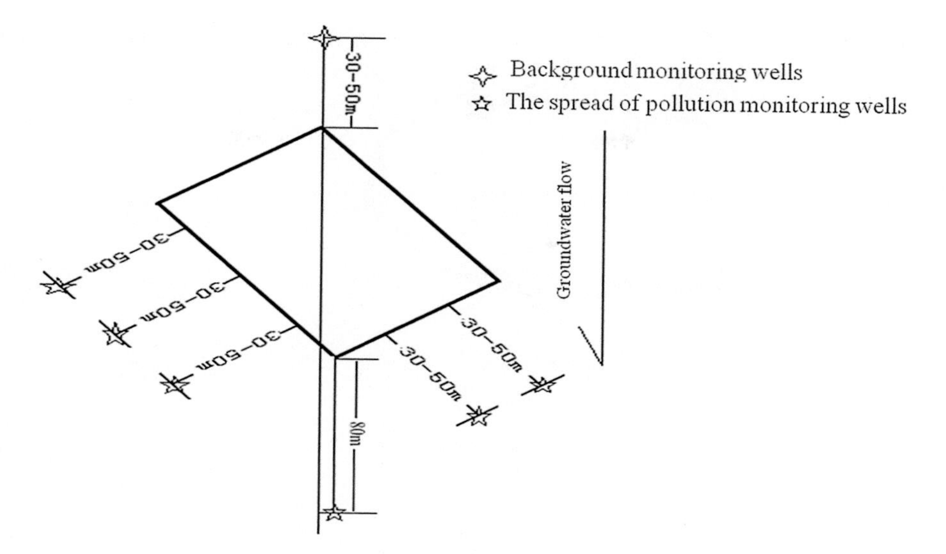

Fig. 5.15 Layout diagram of monitoring wells in the landfill

points closely related to water power of the landfill in upstream direction of the groundwater flow, they can be used as the background monitoring points; 5–6 spread of pollution monitoring wells shall be set, which can be subject to liner, "T" shaped (Fig. 5.17) or "cross" shaped (Fig. 5.18) layout; the monitoring points subject to linear layout can be set along direction of the groundwater flow from mountainous area in the landfill, with an equal spacing of 50–80 m; if there are flow out points of groundwater in the downstream, they can be used as the spread of pollution monitoring points.

5.2.2.2 Setting of Monitoring Points

The setting of monitoring points mainly includes several parts: structure design, drilling, well formation, and well completion. There of structure design of the monitoring wells is the basic work in the preliminary stage, and also an important guarantee for smooth well drilling, depollution during drilling, and cost control of the monitoring wells. The monitoring wells (holes) are used to collect groundwater samples in hazardous waste disposal sites, so as to obtain groundwater stage data. Based on hydrogeological conditions and rock (soil) characteristics of the site, design of the monitoring holes shall be performed on the premise of ensuring no effect on groundwater quality. To grasp chemical properties of pollutants and structure of rock (soil) strata in the site, it is required to pay special attention to selection of drilling technology and well formation materials.

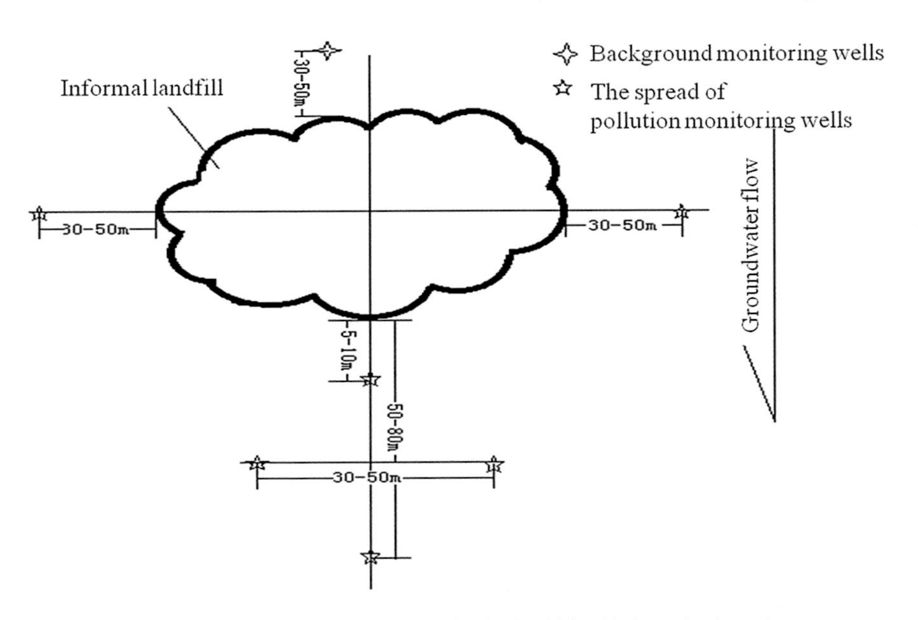

Fig. 5.16 Layout diagram of monitoring wells in the landfill with irregular boundary

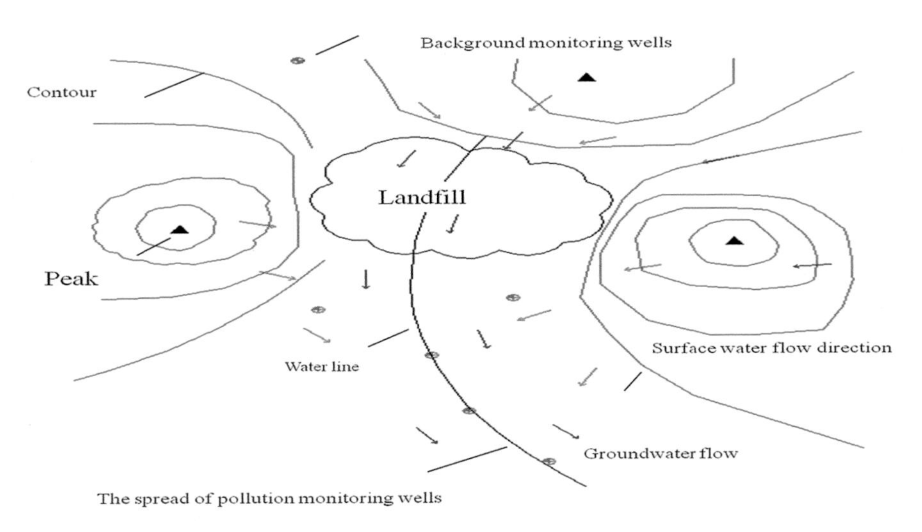

Fig. 5.17 "T" shaped layout diagram

(1) Wellhole

The diameter of the monitoring hole is generally determined by size of the groundwater sampling equipment (e.g., pail, water pump, etc.). In high permeability rock strata, the aquifer is able to provide sufficient groundwater. However, if

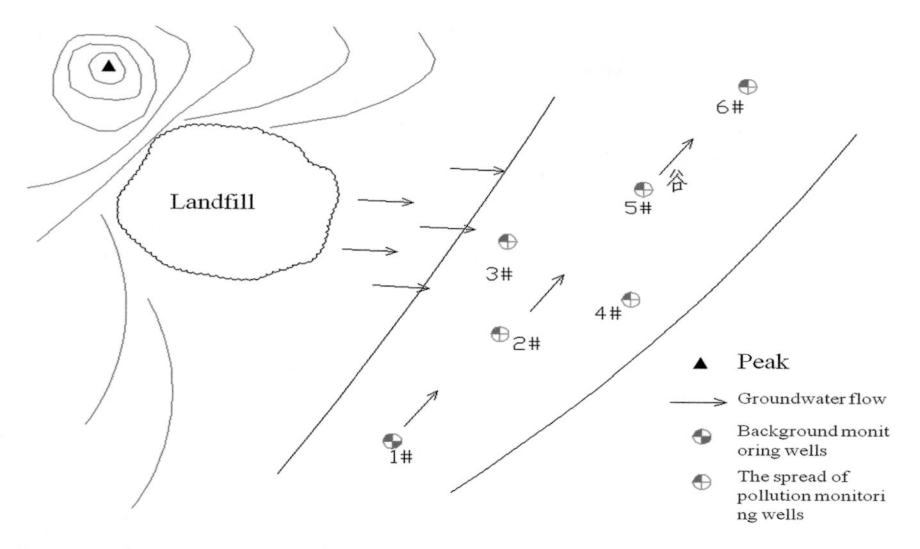

Fig. 5.18 "T" shaped layout diagram

monitoring holes are to be set in areas subject to serious water shortage, even if the hole diameter is large enough, there will be water shortage in extraction of groundwater from the low permeability rock strata. If the groundwater is polluted by noxious liquid waste, large-diameter holes are required for extraction of the groundwater for treatment. Therefore, from the prospective of both safety and treatment cost, it is necessary to minimize extraction of the groundwater in the monitoring stage. For the above reasons, the standard well diameter specified in the technical standard for well formation of monitoring holes is generally 50 mm. If it is also required to proceed to treatment of the groundwater and polluted soil after monitoring, the monitoring hole of large diameter can be used as pumping hole to pump the polluted groundwater for treatment. Besides, the large-diameter wells feature high intensity, so they are often used for deep well monitoring.

(2) Bushing and filter materials

Type of the well formation materials for the monitoring holes has significant effect on quality of the water samples collected. Therefore, the well formation materials shall neither absorb nor filter chemical composition in the water samples, so as to ensure no effect on representativeness of the water samples. The well formation materials generally used are as follows.

(i) Polyvinyl chloride

The polyvinyl chloride (PVC) materials are very cheap and easy to handle, so they are widely used in manufacturing of casings and well filters. PVC is chemically nonreactive in general environments. However, when PVC directly contacts low-molecular-weight ketone, aldehyde, or chloride solutions, it will be subject to denaturalization. Generally, when organic content in the solution increases, it will lead to direct damage or absorption to PVC. Therefore, for eastern coastal areas, especially the hazardous waste disposal sites subject to salinization of ground soil–water environment caused by seawater intrusion, such materials shall not be used for monitoring wells.

(ii) Polypropylene random copolymer

This product features good toughness, high strength, excellent impact resistance, good creep resistance under high temperature, and unique high transparency.

(iii) Polytetrafluoroethylene

Polytetrafluoroethylene is considered as the most chemically inactive well formation material. But because of its high cost, it is only used in the cases where no chemical disturbance is allowed.

(iv) Electroplated casing

The electroplated casing has better performance than the PVC materials, because it is inert to organic compounds and more durable in rock strata. Electroplated film of the electroplated casing also can prevent rusting. But it shall be noted that the electroplated casing will increase concentrations of such elements as iron, manganese, zinc, cadmium, etc. in the groundwater. Concentration of pollutants in the water samples may also be increased as concentration of iron and manganese increases. Therefore, it is more appropriate to use PVC as well formation materials in monitoring of heavy metal pollution in the groundwater.

(v) Stainless steel casing

The stainless steel casing is essentially inert to all pollutants. However, in case of extremely low pH, stainless steel will release chromium ion into the groundwater. This will have a catalytic effect on biodegradation of some organic pollutants. High price is another significant disadvantage of stainless steel casing.

(3) Sealing materials

If rotary or auger drilling method is applied, diameter of the drilled hole shall be greater than that of casing of the monitoring holes, and water-stopping materials (expansive soil slurry, cement slurry or mixture of bentonite and cement, etc.) shall be filled between casing of the monitoring holes and the drill hole wall. Below we will discuss about precautions for use of each filling material.

(i) Bentonite slurry

The bentonite slurry is generally used as the drilling slurry, or as drill hole sealing material after well formation. Structure of the bentonite is formed by Al–Si bonds linked via cationic bridge. Bentonite has strong iron ion exchange capacity, and when approaching inlet of the filter or monitoring hole as a sealing material, chemical composition of the water samples collected may change. It is worth mentioning that, in areas where it is dry and rainless, if bentonite is used as well sealing material, it is likely to cause collapse around the monitoring hole, which is unfavorable for long-term preservation of the monitoring well.

(ii) Cement slurry

After drilling using rotary drilling method and casing running, cement slurry will be used for sealing annular sleeves. For groundwater, cement has higher permeability than bentonite, so cement is sometimes not considered a suitable filling material. While cement is a rigid material, and it can easily form into a whole around casing of the monitoring hole. It should be noted that improper use of cement may affect pH of the water samples.

(iii) Mixture of bentonite and cement

The mixture of bentonite and cement is often used as filling slurry. The mixed slurry has slightly lower strength and higher permeability compared with the straight cement. Variation of the mixture helps to strengthen structural strength and permeability resistance of the filling slurry.

(iv) Length and embedding depth of filter

Length of filter of the monitoring hole and its embedding depth under the ground are determined by nature of the pollutants in the aquifer and vadose zone and the monitoring purposes. When the aquifer used as a water supply source is to be monitored, filters shall be installed throughout thickness range of the aquifer. However, if it is required to take samples within a specific depth interval, usually multiple vertical monitoring points will be used, viz., the depth-specific sampling mode. This technology is also very necessary if aquifer of the groundwater is too thick for monitoring even long filter is used.

It calls for special attention that the light nonaqueous phase liquids, viz., the liquid pollutants that are less dense than water, will float above the groundwater surface. In monitoring of such floating pollutants, length of the filter must be extended to the entire saturated zone of groundwater, so that these light liquids can enter the monitoring holes. Length and location of the filter must correspond to the groundwater stage and its corresponding change.

5.2.2.3 Sampling

(1) Sampling frequency

For the monitoring wells, sampling shall be conducted once every quarter, for totally 4 times; if additional monitoring wells are drilled as the monitoring wells in the landfill fail to meet the quantity and quality requirements, respectively, one soil sample can be taken from each soil stratum along vertical depth from the ground surface to the groundwater stage for analysis and detection, only for the new monitoring wells. If leachate can be collected, the monitoring frequency shall be the same with that of the groundwater, namely, once every quarter, for totally 4 times in the whole year.

(2) Preparation for sampling

Before sampling, one director shall be assigned for organization and implementation of the investigation in each solid waste disposal site. The sampling director shall make clear of the investigation purposes and requirements, get familiar with the groundwater sampling procedures and methods, formulate detailed sampling schedule (subject to demonstration by experts with rich experience in groundwater on-site sampling), implement rational division of labor for the on-site sampling, and make sampling plan. The sampling plan shall include the following items: sampling purpose, monitoring well location, monitoring items, sampling quantity, sampling time and route, sampling quality assurance measures, sampling equipment and vehicles, items requiring field monitoring, personnel safety guarantee and clear division of labor for the sampling personnel, etc., and training for on-site sampling shall be organized if necessary. Meanwhile, it is also required to supervise the on-site sampling personnel to implement sampling in strict accordance with relevant technical specifications.

Besides, purchase and allocation of the on-site sampling equipment is also an important part for the preliminary preparation. The sampling equipment shall meet the requirements on the samplers and water sample containers in *Technical Specifications for Environmental Monitoring of Groundwater* (HJ/T 164-2004). The necessary instruments and equipment in the site shall be provided with detailed list, and checked and packed by the sampling director one by one, so as to avoid mistake or missing.

(3) Sampling methods and requirements

For well water requiring stage measurement, the groundwater stage shall be measured before sampling; water sampling from the well must be carried out after full swabbing when coefficient of variation of three consecutive conductivity measurements is less than 3 %. The necessary water sampling amount for each monitoring item shall meet the requirements in *Technical Specifications for Environmental Monitoring of Groundwater* (HJ/T 164-2004). After sampling, the water sample containers shall be tightly capped and sealed immediately, and reliably labeled, and the label design shall be determined according to actual situations, generally including such information as number of the monitoring wells, sampling date and time, monitoring items, sampling person, etc. Before completion of sampling, it is required to check the sampling plan, sampling records, and water samples, and in case of any mistake or missing, resampling or supplementary sampling shall be conducted immediately.

(4) On-site monitoring items

For groundwater in the built landfills, all the items available for on-site measurement shall be measured on the site, including eight indexes: water temperature, pH, redox potential, conductivity, turbidity, color, smell, and taste. Besides, it is also

Table 5.3 Groundwater on-site sampling record

Number of the monitoring well	Monitoring well	Sampling date			Sampling time	Sampling method	Sampling depth (m)	Tempera ture (°C)	Weather conditions
	Name	Year	month	day					

On-site measurement records										
Stage (m)	Flow (m^3/s)	Temperature (°C)	Color	Redox potential	Smell and taste	Turbidity	Visible material by bare eye	pH	Conductivity (μs/cm)	Properties of the sample
Remarks										

Sampler Recorder Date

Table 5.4 Monitoring items for groundwater and leachate in the landfills

Index type			Index name	Index number
Inorganic	Required		Total dissolved solids, total hardness, permanganate index, nitrate nitrogen, ammonium nitrogen, nitrite nitrogen, sulfate, carbonate, chloride ion, fluoride ion, iodine ion, sodium, potassium, calcium, magnesium, iron, manganese, lead, zinc, cadmium, hexavalent chromium, mercury, arsenic, selenium, total coliform group, and total bacterial count	25
	Optional*		Cyanide, sulfide, total phosphorus, bromine, total chromium, copper, barium, beryllium, molybdenum, nickel, boron, antimony, silver, and thallium, or total α and β radioactivity, which can be added according to the proportion of 10–20 % of the total quantity	16
Organic	Required	Halohydrocarbons	Trichloromethane, tetrachloromethane, bromodichloromethane, tribromomethane, and chloroethylene	16
		Calorinatedbenzenes	Chlorobenzene	
		Monocyclic aromatic hydrocarbons	Benzene, methylbenzene, ethylbenzene, dimethylbenzene, and phenylethylene	
		Organochlorine pesticides	Total benzene hexachloride, dichloro-diphenyl-trichloroethane, p′-DDT, and hexachlorobenzene	
		Polycyclic aromatic hydrocarbons	Benzo (a) pyrene	
	Optional*	Comprehensive indexes	TVOC, TOC	12
		Phenols	Pentachlorophenol, m-cresol, phenol, and p-Nitrophenol	
		Polychlorinated biphenyls	Polychlorinated biphenyl	
		Others	Dichloroacetic acid, trichloroacetic acid, trichloroacetaldehyde, nitrobenzene, and aniline	

Notes 1. For the "optional" indexes, particular pollutants shall be screened for monitoring according to type of the landfill waste, and for optional inorganic and organic indexes, at least three items shall be selected, respectively, for monitoring in formal municipal solid waste landfills, and five items for that in informal municipal solid waste landfills required to determine the temperature, describe the weather conditions and recent rainfall, and truthfully fill in the groundwater on-site sampling record, viz., Table 5.3.

5.2.2.4 Monitoring Items

Main screening principles for monitoring items of groundwater in the landfills are as follows: (1) Select the monitoring items to be controlled as specified in *Quality Standard for Ground Water* (GB/T 14848), so as to meet the requirements for quality assessment and protection of groundwater. (2) Appropriately add some optional monitoring items according to category of incoming waste and function of the groundwater in the landfill. (3) Provide national or industrial standard analysis method, industrial technical specifications for monitoring, and industrial unified analysis method for the selected monitoring items. Monitoring items of groundwater are consistent with that of the leachate, as shown in Table 5.4.

5.3 Optimization and Application of 3MRA Model

5.3.1 Introduction of the Model System

3MRA (multimedia, multi-pathway, and multi-receptor exposure and risk assessment) model, developed by the US Environmental Protection Agency in 1999, was originally used to estimate the potential risk to human health and ecological environment during treatment and disposal of dangerous wastes using nonhazardous waste disposal unit, and further determine whether to implement exemption of the hazardous waste from regulatory control according to the calculated risk value. It is now also used for assessment of potential human health and ecological risk due to long-term (chronic) exposure to the specified wastes. The 3MRA model consists of 17 interrelated sub-models, including five waste disposal unit models (also called "source release model"), five medium models (also called "migration and trending/transformation model"), three food chain models, and four exposure and risk characterization models (Babendreier and Castleton 2005; USEPA 1999a, b, 2003) (Fig. 5.19).

5.3.2 Risk Assessment of Groundwater in China's Landfill Areas Based on the 3MRA Model

5.3.2.1 Construction of Simulation Domain

Before operation of the model, it is required to first construct the simulation domain corresponding to the survey region. Select similar sites from site database of the

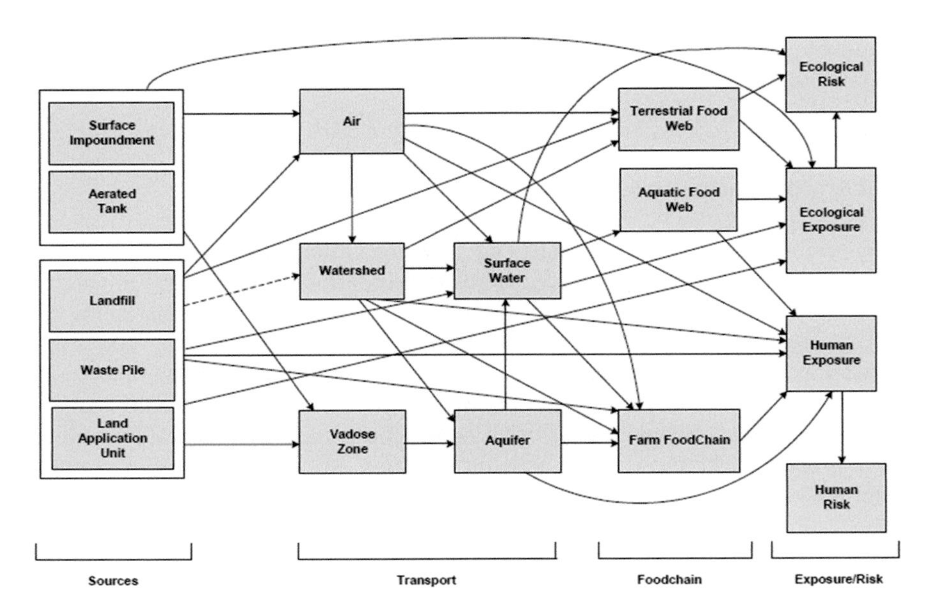

Fig. 5.19 Framework of 3MRA model

model as representative sites according to such parameters as hydrogeological characteristics, climate characteristics, crowd distribution characteristics, etc., in the survey region, and ensure that range of the conditions in the survey region is included in the domain constructed by the selected representative sites, viz. survey region $\in$ simulation domain, so as to ensure that the simulation domain can fully reflect the characteristic parameters in the survey region, make operation results of the model more representative and reliable, and improve authenticity of the simulation.

5.3.2.2 Model Simulation

Select seven pollutants of hexavalent chromium (Cr^{+6}), lead (Pb), bivalent nickel (Ni^{+2}), arsenic (As), mercury (Hg), cadmium (Cd), and benzene for risk assessment of groundwater in 17 provincial and municipal solid landfill areas in China. Select the sites contained in the model domain as the site input data; set totally 1–5 waste level(s); and implement simulation by corresponding parameter conditions such as 11,031 iterative calculations using the dual Monte Carlo model. The model will generate seven forms under five conventional standards by default. Conditions and limits of the five conventional standards are as shown in Table 5.5 (USEPA 2003).

Adduct characterization method shall be used to calculate risk and hazard index (hazard quotient/noncarcinogenic risk) of the groundwater in the landfill areas. First, assuming that risk of water intake, shower and pure groundwater pathways are, respectively, variables x_1, x_2, and x_3, and total risk of the groundwater is the variable R, then the following formula can be established:

$$R = \sum_{n=1}^{3} x_n \tag{5.3.1}$$

Second, for each pathway, the final risk value is adduct of the risk value of all the pollutants. Assuming that corresponding risk value of the seven pollutants of benzene, As, Cd, Cr^{+6}, Hg, Ni^{+2}, and Pb are, respectively, the variables $y_1 \sim y_7$, then the following formula can be established:

$$x_n = \sum_{m=1}^{7} y_m \tag{5.3.2}$$

Based on the above two formulas, calculation formula for the total risk of groundwater in the landfill area can be obtained:

$$R = \sum_{n=1}^{3} \sum_{m=1}^{7} y_m \tag{5.3.3}$$

Based on the same calculation method, assuming that hazard index of the pollutants is the viable Z, then calculation formula for the hazard index HQ of groundwater in the landfill area can be obtained as follows:

$$HQ = \sum_{n=1}^{3} \sum_{m=1}^{7} z_m \tag{5.3.4}$$

Operation results of the model can be processed based on the above methods to obtain carcinogenic risk and hazard index of the groundwater in the 17 landfill areas.

Table 5.5 Conditions and limits of the conventional standards

Standard	Standard I	Standard II	Standard III	Standard IV	Standard V
Risk level	1.00E−06	1.00E−06	1.00E−05	1.00E−05	1.00E−07
Hazard index of human receptors (hazard quotient)	1	1	1	10	10
Hazard index of ecological receptors (hazard quotient)	1	1	1	10	10
Crowd protection weight (%)	99	99	99	95	95
Protection possibility (%)	95	85	85	85	90

5.3.2.3 Model Export

The model production is based on the corresponding risk value for each exposure pathway in 17 landfill areas under five kinds of conventional criteria and landfill (LF) modes. Through aforesaid sum representation method, water ingestion pathway, shower pathway, and groundwater pathway are selected to jointly represent the carcinogenic risk and noncarcinogenic (hazard index) risk of the groundwater in landfill area.

As shown in Table 5.6, the risk level in 17 landfills varies differently in different landfills under different criteria. Among the five criteria, the highest frequency for the highest carcinogenic risk and hazard index occurs in landfills under Criteria III with 11 times, which is followed by Criteria II. Criteria I and Criteria V rank third in the list, and Criteria IV ranks last. Such data indicates that Criteria III can reflect

Table 5.6 Frequency distribution for the highest carcinogenic risk and hazard index of groundwater in landfill area

Landfill	Standard I	Standard II	Standard III	Standard IV	Standard V	Total
Jilin Province	0	0	2	1	1	4
Qingdao, Shandong	1	1	0	0	0	2
The East Part of Inner Mongolia	1	2	2	1	1	7
Beijing	0	0	0	1	2	3
Taiyuan, Shanxi	1	0	0	0	0	1
Hefei, Anhui	3	1	3	2	2	11
Guangdong Province	0	0	0	0	0	0
Dalian	1	1	2	1	1	6
Heilongjiang Province	1	2	2	1	2	8
Changsha, Hunan	0	0	0	0	0	0
Changshou, Chongqing	0	0	0	0	0	0
Jiangxi Province	1	1	0	0	0	2
Fujian Province	0	0	0	0	0	0
Hanzhong, Shaanxi	0	0	0	0	0	0
Xinjiang Uygur Autonomous Region	0	0	0	0	0	0
Anyang, Henan	0	1	0	0	0	1
Hubei Province	0	1	0	0	0	1
Total value	9	10	11	7	9	

the fragility size and character of particular pollutants (benzene, arsenic, cadmium, mercury, divalent nickel, and hexavalent chromium) in waste landfills in China more comprehensively and detailedly, and reflect the carcinogenic risk and hazard property of landfill wastes in China more synthetically and explicitly. It will provide direction guidance to find out the existed common problems in construction, operation, supervision, risk pre-warning, pollution emergency, and other aspects for waste landfills in China, and provide guidance suggestions for management operation of landfills in China.

On the basis of data in Table 5.6, the frequency distribution for the highest carcinogenic risk and hazard index of groundwater in landfill area as shown in Fig. 5.20 shall be obtained through GIS platform.

In 17 landfills, the carcinogenic risk and hazard index of groundwater in landfill area is relatively higher in the Northeast, North China, and central China. The standard exceeding phenomenon occurs under five criteria. The groundwater system is relatively fragile in some areas, and is more vulnerable to pollution from pollutant in landfill wastes, so it shows higher environmental risk. The groundwater in South China and Southwest region shows relatively modest carcinogenic risk, which is generally one order of magnitude lower than standard value at safety level. The carcinogenic risk and hazard index is relatively lower for the groundwater in landfill area in the Northwest, which is generally 2–3 orders of magnitude lower, and the groundwater in the area shows better pollution-proof performance and lower fragility.

Based on the above conclusions, it is suggested that zoning classification control shall be implemented for wastes in China. It is suggested that nearby landfill shall be selected to propose the wastes at lower risk level, so as to reduce the disposal

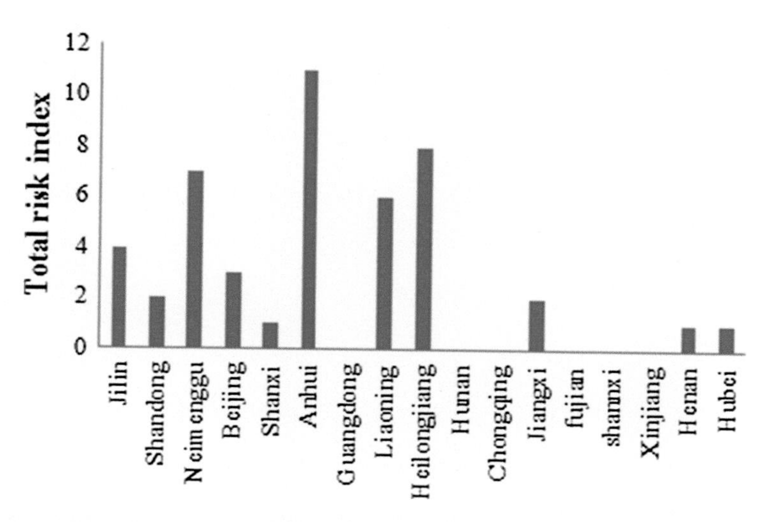

Fig. 5.20 Frequency distribution diagram for the highest carcinogenic risk and hazard index of groundwater in landfill area

costs for such wastes; for wastes at higher risk level, it is suggested that detoxification pretreatment shall be conducted, so as to implement dumping after the environmental risk has been reduced. It is also considered that the wastes at higher risk level shall be transported to landfill areas with lower groundwater fragility and better ecological stability, so as to realize hierarchical control for wastes, and improve the wastes disposal efficiency.

References

Babendreier JE, Castleton KJ. Investigating uncertainty and sensitivity in integrated, multimedia environmental models: tools for FRAMES-3MRA [J]. Environ Model Softw. 2005;20(8): 1043–55.

Khanbilvardi RM, Ahmed S, Gleason PJ. Flow investigation for landfill leachate (FILL) [J]. J Environ Eng. 1995;121:45–57.

USEPA. Human exposure module for HWIR99 multimedia, multipathway, and multireceptor risk assessment (3MRA) Model. Washington: Office of Solid Waste; 1999a.

USEPA. Source Modules for tanks and surface impoundments: Background and implementation for the multimedia, multipathway, and multireceptor risk assessment (3MRA) for HWIR99. Washington: Office of Solid Waste; 1999b.

USEPA. Multimedia, multipathway, and multireceptor risk assessment (3MRA) modeling system volume I: Modeling system and science [R]. Athens: GA and Research Triangle Park; 2003. P. 1–6.

Chapter 6
Ranking Management Technology System for Groundwater Pollution Risk of Landfill

Abstract To realize scientific hierarchy and effective management for groundwater pollution of landfill in China, and to ensure that the landfills with significant occurrence probability of groundwater pollution accident are subjected to major supervision, index system and technical methods suitable for groundwater pollution risk ranking of landfill in China are established in this chapter. Management procedures and methods for groundwater pollution of landfill are proposed on this basis. In combination with risk control node, specific risk control measures and management plan are proposed in respect to the established landfills and landfills under planning and construction respectively. It provides scientific basis and technical support for the environmental protection department in China to carry out ranking management for groundwater pollution risk of landfill, and to ensure environmental safety of groundwater.

Keywords Landfill · Groundwater · Risk ranking · Ranking management

6.1 Index System for Groundwater Pollution Risk Ranking of Landfill

6.1.1 Establishment Principles for Index System

In the risk evaluation index system for groundwater pollution of landfill, the natural attribute characters of aquifer and the inherent characters of landfill shall be considered. On the basis of the process that pollution risk for groundwater is caused by landfill, its risk evaluation index system shall be established step-by-step. While selecting the index for construction of index system, the following principles must be followed:

B. Xi et al., *Optimization of Solid Waste Conversion Process and Risk Control of Groundwater Pollution*, SpringerBriefs in Environmental Science,
DOI 10.1007/978-3-662-49462-2_6

1. Principle of scientificity: the index system shall be established on the basis of scientific theory. The specific index can objectively and truly reflect the correlation among various influence factors, as well as the internal mechanism in landfill that caused groundwater pollution risk. Meanwhile, the concept of each index must be clear, and the statistical measurement method must be normative. In this way, the comparability of index can be ensured, and the scientificity of evaluation methods, as well as authenticity and objectivity of evaluation results can be ensured.
2. Principle of comprehensiveness and independence.
3. Principle of quantification and feasibility: the established index system always has better theoretical reflection, but lacks practicality. Therefore, it shall not be divorced from reality of data and information conditions related to the index during index selection. The critical indexes with comprehensiveness shall be selected as far as possible, so that the established index system is concise, and easy for calculation and analysis.

6.1.2 Methods and Procedures for Construction of Index System

Using system analysis concept, complicated problems are divided into several associated and ordered hierarchies by analytic hierarchy process (Chen et al. 2005), and the coherent elements of each hierarchy are compared and analyzed. In general, there are three layers: destination layer, criterion layer, and index layer. In the event of relatively complicated problems, each layer can also be subdivided into secondary layers, so as to further analyze the problems.

Destination layer (A), namely overall objective of groundwater pollution risk ranking, is the highest layer of index system.

Criterion layer (B) is the major system layer to ensure realization of overall objective.

Index layer (C) refers to the most basic layer of index system, which includes all specific indexes of risk ranking. These indexes are direct measurable factors to evaluate the risk level and environmental impact of groundwater pollution in hazardous waste disposal sites.

On the basis of the aforesaid methods for index establishment in the chapter, and according to the analysis of process that risk for groundwater is caused by solid wastes dumping, the natural attribute characters of aquifer and inherent characters of landfill are comprehensively considered, and the index system for pollution risk ranking groundwater in hazardous wastes landfill shall be analyzed and established step-by-step.

6.1.3 Construction of Index System

Referring to the selection method for ranking index in Canada's pollution risk ranking method (Canadian Council of Ministers of the Environment 2008), it is confirmed that the selection for risk ranking index of groundwater pollution of landfill in China includes the following three aspects according to the establishment procedure and principles to follow for ranking index: (1) the risk indexes existed in landfill, which represents the risk of leachate entering vadose zone across impermeable membrane after the leakage of impermeable membrane in landfill; (2) the vertical migration risk index of groundwater pollutant in landfill, which represents the risk of the pollutant entering aquifer through vadose zone; (3) the horizontal migration risk index of groundwater pollutant in landfill, which represents the risk of the pollutant's relocation diffusion to drinking water source after entering aquifer.

The risk existed in landfill mainly refers to the risk of groundwater pollution caused by the inherent conditions and properties of landfill, which is composed of the scale of landfill construction (including landfill capacity per year, area of landfill, life of landfill, etc.), as well as the characters of leachate and seepage control system (leachate output, thickness of HDPE impermeable membrane in landfill, and thickness of clayliner).

The vertical migration risk of groundwater pollutant in landfill mainly reflects the risk of pollutant entering aquifer through soil-vadose zone media, which is mainly determined by the hydrological condition and the nature of vadose zone media. The hydrological conditions include depth of groundwater, net recharge of regional groundwater, and terrain slope, and the nature of vadose zone media includes thickness and permeability coefficient of vadose zone.

The horizontal migration risk of groundwater pollutant in landfill mainly reflects the risk of pollutant's relocation diffusion to drinking water source with groundwater flow after entering aquifer, which is mainly determined by the nature of aquifer media (thickness and permeability coefficient of the aquifer) and the distance between drinking water source and landfill.

6.1.4 Index Weight Assignment

Since 14 indexes in three aspects (including the inherent risks in landfill, the vertical migration risk of groundwater pollutant in landfill and the horizontal migration risk of groundwater pollutant in landfill) are included in the pollution risks of landfill on groundwater, so the pollution risks of landfill on groundwater shall be represented by comprehensive risk index, and different weight assignments for indexes representing risks is the key to obtain the comprehensive risk index. In the paper, analytic hierarchy process is used to calculate the total target weight of each index.

Table 6.1 Hierarchical structure for groundwater pollution risk ranking index of landfill landfill

Destination layer A	Criterion layer C	Criterion layer W
Groundwater pollution risk ranking index of landfill	Inherent risk index of landfill C1	Landfill capacity per year W11
		Area of landfill W12
		Life of landfill W13
		Leachate output W14
		Thickness of HDPE impermeable membrane W15
		Thickness of claylinerW16
	Vertical migration risk index of pollutant C2	Depth of groundwater W21
		Net recharge of groundwater W22
		Terrain slope W23
		Thickness of vadose zone W24
		Permeability coefficient of vadose zone W25
	Horizontal migration risk index of pollutant C3	Thickness of the aquifer W31
		Permeability of the aquifer W32
		Distance between drinking water source and landfill W33

Step I: Establish hierarchical structure model
The hierarchical structure model is established according to the index system for groundwater pollution risk ranking of landfill (refer to Table 6.1).

Step II: Using paired-comparisons method, confirm the judgment matrix between destination layer A and criterion layer C and the judgment matrix between criterion layer C and criterion W, respectively. Tables 6.2, 6.3, 6.4 and 6.5.

Step III: Calculate the maximum eigenvalue of A ~ C and C ~ W judgment matrix and its corresponding eigenvector:

Maximum eigenvalue of A ~ C Judgment matrix: $\lambda max = 3$,
Eigenvector: $W = (0.333, 0.333, 0.334)$ T
Consistency check: $CR = CI/RI = 0/0.58 = 0 < 0.1$
Maximum eigenvalue of C1 ~ W Judgment matrix: $\lambda max = 6.009$,
Eigenvector: $W = (0.039, 0.079, 0.118, 0.195, 0.300, 0.269)$ T
Consistency check: $CR = CI/RI = 0.009/1.24 = 0.007 < 0.1$

Table 6.2 A ~ C judgment matrix

A	C1	C2	C3
C1	1	1	1
C2	1	1	1
C3	1	1	1

Table 6.3 C1 ~ W judgment matrix

C1	W11	W12	W13	W14	W15	W16
W11	1	1/2	1/3	1/5	1/8	1/7
W12	2	1	3/5	3/7	1/4	1/3
W13	3	5/3	1	4/7	3/8	4/9
W14	5	7/3	7/4	1	2/3	5/7
W15	8	4	8/3	3/2	1	1
W16	7	3	9/4	7/5	1	1

Table 6.4 C2 ~ W judgment matrix

C2	W21	W22	W23	W24	W25
W21	1	9/2	5	2	1
W22	2/9	1	7/8	2/5	2/9
W23	1/5	8/7	1	2/5	1/5
W24	1/2	5/2	5/2	1	1/2
W25	1	9/2	5	2	1

Table 6.5 C3 ~ W judgment matrix

C3	W31	W32	W33
W31	1	1/2	5
W32	2	1	9
W33	1/5	1/9	1

Maximum eigenvalue of C2 ~ W Judgment matrix: $\lambda max = 5.006$,
Eigenvector: W = (0.341, 0.071, 0.072, 0.175, 0.341) T
Consistency check: CR = CI/RI = 0.006/1.12 = 0.005 < 0.1
Maximum eigenvalue of C3 ~ W Judgment matrix: $\lambda max = 3.001$,
Eigenvector: W = (0.319, 0.615, 0.066) T
Consistency check: CR = CI/RI = 0.001/0.58 = 0.002 < 0.1.

Step IV: Refer to Table 6.6 for overall hierarchical ranking calculation and weight calculation results of pollution risk ranking index of groundwater in hazard wastes landfill:

Table 6.6 Weight calculation results of groundwater pollution risk ranking index of landfill

Index	W11	W12	W13	W14	W15	W16	W21
Weigh	0.013	0.026	0.039	0.065	0.1	0.09	0.114
Index	W22	W23	W24	W25	W31	W32	W33
Weigh	0.024	0.024	0.058	0.114	0.106	0.205	0.022

$$W = \begin{pmatrix} 0.039 & 0 & 0 \\ 0.079 & 0 & 0 \\ 0.118 & 0 & 0 \\ 0.195 & 0 & 0 \\ 0.300 & 0 & 0 \\ 0.269 & 0 & 0 \\ 0 & 0.341 & 0 \\ 0 & 0.071 & 0 \\ 0 & 0.072 & 0 \\ 0 & 0.175 & 0 \\ 0 & 0.341 & 0 \\ 0 & 0 & 0.319 \\ 0 & 0 & 0.615 \\ 0 & 0 & 0.066 \end{pmatrix} \cdot \begin{pmatrix} 0.333 \\ 0.333 \\ 0.334 \end{pmatrix} = \begin{pmatrix} 0.013 \\ 0.026 \\ 0.039 \\ 0.065 \\ 0.1 \\ 0.09 \\ 0.114 \\ 0.024 \\ 0.024 \\ 0.058 \\ 0.114 \\ 0.106 \\ 0.205 \\ 0.022 \end{pmatrix}$$

6.2 Technical Methods for Groundwater Pollution Risk Ranking of Landfill

6.2.1 Risk Ranking Methods

Step I: Primary screening for groundwater pollution risk

As for the landfill that has been established and put into operation, it shall be first determined whether any items in the particular pollutants of groundwater have exceed the standard of Grade III in *Quality Standard for Ground Water* (GB14848–93). If so, its groundwater pollution risk shall be divided into Grade I directly, and shall be subjected to major supervision; otherwise, proceed to the next ranking procedures.

As for the landfill that has not been established and put into operation, conduct the next ranking procedure directly.

Step II: Index quantification for groundwater pollution risk ranking of landfill

On the basis of the conducted primary screening for groundwater pollution risk of landfill, the basic data of landfill shall be collected and organized in detail in the manner and basic data collection and site survey, including the contents of three aspects, namely the inherent risk existed in landfill, the vertical migration risk of groundwater pollutant in landfill, and the horizontal migration risk of groundwater pollutant in landfill. The various indexes shall be quantized as shown in Table 6.7.

Step III: Determination of landfill risk index

On the basis of basic site data collected in Table 6.7, each index is analyzed with the numerical analysis method of cluster analysis. According to the results of cluster analysis, and in combination with the numeric feature of each index and the relationship between index value

Table 6.7 Basic data collection for landfill

Inherent risk index of landfill			
Scale of landfill	Landfill capacity (t/a)	Area of landfill (m^2)	Life of landfill (a)
Leachate generation and seepage-proofing	Leachate output (m^3/d)		
	Thickness of impermeable membrane (HDPE membrane) (mm)		
	Thickness of clayliner (mm)		
Vertical migration risk index of groundwater pollutant in landfill			
Hydrological condition	Depth of groundwater (m)	Net recharge of groundwater (mm)	Terrain slope (%)
Nature of vadose zone media	Thickness of vadose zone (mm)	Permeability coefficient of vadose zone (cm/s)	
Horizontal migration risk index of groundwater pollutant in landfill			
Nature of the aquifer media	Thickness of the aquifer (mm)	Permeability of the aquifer (cm/s)	
Distance between drinking water source and landfill (m)			

and risk value, ranking for each index is conducted, and ranking boundary is given. The total target weight of each index is calculated with analytic hierarchy process. On this basis, the comprehensive rating for groundwater pollution risk of landfill is conducted with weighting method.

Step IV: Landfill risk ranking
Using cluster analysis method, the comprehensive rating and analysis for groundwater pollution risk of landfill are conducted, and the ranking results and ranking boundary of underground pollution risk in landfill are obtained as a result.

6.2.2 *Case Analysis*

The chapter ranks the groundwater contamination risk and gives the ranking boundary through established risk ranking method and screened risk ranking index on the basis of the data obtained from site survey and collected from basic information on China's 37 hazardous waste landfills.

First, the groundwater contamination risk is preliminarily screened.

The hazardous waste landfills that have been constructed and put into operation in China are provided with site survey. The groundwater is sampled and monitored. The monitoring data show that all the particular pollutants in the surrounding groundwater do not exceed the three-level criteria of *Quality Standard for Groundwater* (GB14848–93). Therefore, the next ranking procedure starts.

Second, the risk ranking index of groundwater contamination in the hazardous waste landfill is quantized.

The basic information on China's 37 hazardous waste landfills are collected and organized based on 14 indexes listed in Table 6.7. See Table 6.8.

Third, the risk index of hazardous waste landfill is determined.

I. The index ranking boundary is determined with cluster analysis method.
14 indexes listed in Table 6.7 can be divided into two categories. The first category is positively correlated with risk. The greater the value is, the larger the risk that the index results in it will be, such as amount of landfill, area of landfill, life of landfill, amount of leachate, amount of net recharge, terrain slope, permeability coefficient of vadose zone, permeability coefficient of aquifer, and thickness of aquifer. The second category is negatively correlated with risk. The greater the value is, the smaller the risk that the index results in it will be, such as depth of groundwater, thickness of HDPE membrane, distance, thickness of clay liner, and thickness of vadose zone. The chapter introduces the ranking boundary classification method for two categories of indexes by taking the examples of depth of aquifer and depth of groundwater.

Table 6.8 Statistics of groundwater contamination risk index in China's hazardous waste landfills

No.	Depth of groundwater (m)	Amount of the landfills (t/a)	Area of landfill (m^2)	Life of landfill (a)	Thickness of HDPE (mm)	Amount of leachate (m^3/d)	Distance (m)
1	0.5	11092.2	17173	20	3.5	4.8	1100
2	9.7	13000	45736	16	3	5	800
3	126	35848	53465	25	3	12	800
4	2.4	18885	36200	16	3.5	6	800
5	7.5	33395	200000	20	3.5	27.5	2100
6	2.5	70000	51200	29	3	11.8	1200
7	5.3	10000	76000	22	3	50	1000
8	19.5	21984.6	37200	11	3.5	30	800
9	20	10458	115800	15	3	38	800
10	40	25920	143000	15	3.5	72	800
11	2.5	50950	194000	18	3.5	50	800
12	25	26280	361000	15.5	3.5	23.2	1600
13	10	24180	118400	10	3.5	30	1400
14	1.55	25000	58400	20	3	50	1200
15	4.7	9940	142500	20	3.5	11.33	1300
16	20	40000	133450	25	4	20	1200
17	3.2	16000	72000	12	3	40	1500
18	8	15800	53333.6	22	3	50	800
19	7.97	70805	127334	24	3	12	800
20	3	15000	111800	15	3	10	800

(continued)

Table 6.8 (continued)

No.	Depth of groundwater (m)	Amount of the landfills (t/a)	Area of landfill (m^2)	Life of landfill (a)	Thickness of HDPE (mm)	Amount of leachate (m^3/d)	Distance (m)
21	81.5	60430	44900	20	3.5	2.1	1200
22	3	57600	125700	12	3	30	1020
23	30	40000	61535	10	4	2.5	4660
24	50	82700	161277	20	3.5	42	1200
25	2.4	31800	36800	10	3	5.5	5000
26	50	25100	42000	20	3	5.4	1300
27	5	11700	100000	20	3.5	4.4	2000
28	3.8	10000	34110	30	4	12	950
29	1	7115	9000	20	2.5	0.65	2000
30	50	15500	53000	20	4	3.6	900
31	5	16800	102000.5	20	4	12	800
32	3.8	63195	7200	30	3.5	60	952
33	4.1	40000	110000	8	4	30	1500
34	1	10000	87160	8	3.5	35	2000
35	1.2	21152	32000	14.5	3.5	20	2500
36	13	10000	7200	10	2	0	800
37	1	25000	87160	47.6	2.5	30	800
No.	**Thickness of clay liner (cm)**	**Amount of net recharge (mm)**	**Terrain slope**	**Thickness of the vadose zone (m)**	**Permeability coefficient of the vadose zone (m/d)**	**Thickness of the aquifer (m)**	**Permeability coefficient of the aquifer (m/d)**
1	200	101.3	0.01	5.1	0.035	8.6	2.9
2	30	198.3	0.08	7.65	0.007	3.3	0.014
3	50	168.62	0.15	8.5	0.015	19	0.001
4	60	173.78	0.04	3	0.013	3	0.013
5	50	47.2	0.02	5.6	0.35	7	0.002
6	100	278.92	0.11	4.5	0.001	5	1.5
7	50	174.26	0.08	4.6	0.49	13.5	0.046
8	100	129.04	0.02	6.9	0.003	15	0.003
9	50	233.04	0.05	5	0.01	20	0.2
10	50	208.52	0.09	5.9	0.0377	21.84	0.222
11	80	247.04	0.04	5	6	6	0.016
12	60	135.68	0.06	10	0.03	20	6
13	60	62.18	0.03	6	0.003	5.8	0.014
14	60	131.14	0.07	1.25	0.26	1.31	0.005
15	100	159.92	0.01	1.1	0.01	7.7	0.009
16	60	265.5	0.02	16.1	0.066	16.6	5.219
17	50	91.26	0.02	7.3	0.001	8	0.864
18	50	330.62	0.18	4.5	0.026	16	0.026
19	60	134.84	0.1	1.9	0.029	3.4	30
20	100	273.74	0.2	8.4	0.004	24	0.005
21	100	88.24	0.12	10	0.05	20	0.5

(continued)

Table 6.8 (continued)

No.	Depth of groundwater (m)	Amount of the landfills (t/a)	Area of landfill (m^2)	Life of landfill (a)	Thickness of HDPE (mm)	Amount of leachate (m^3/d)	Distance (m)
22	150	354.37	0.1	5	0.01	9.39	0.06
23	50	137.58	0.05	1.3	0.226	5.2	2.644
24	60	123.22	0.02	2.5	0.003	2.4	0.035
25	90	269	0.15	3.1	0.001	9	0.013
26	50	58.9	0.02	24.7	0.02	8.2	0.127
27	50	51.04	0.03	4.4	0.05	2.9	5
28	50	279	0.06	3.7	0.035	15	0.52
29	50	12.58	0.01	4.6	0.11	4.5	2.03
30	50	110.8	0.03	3	0.002	5	0.02
31	50	199.62	0.1	5.8	0.001	15.2	0.01
32	60	270	0.2	5.5	0.05	5	0.1
33	60	378	0.18	5	0.01	9.39	0.06
34		279.04	0.1	6	0.001	20	0.001
35	60	302.98	0.02	6	0.001	20	0.001
36	30	106.8	0.02	12	0.014	27	1.4
37	50	215.52	0.01	9	0.001	9	0.34

1. Thickness of the aquifer
   With the same hydraulic gradient and flow velocity, the thinner the aquifer where the landfill is located, the smaller the runoff of groundwater flowing over the landfill will be, and the less the spreading effect of hazardous substances will be. It is beneficial for the protection of groundwater. Contrarily, with the same hydraulic gradient and flow velocity, the thicker the aquifer is, the greater the runoff of groundwater flowing over the landfill will be, and the larger the spreading range of hazardous substances will be. It will increase the risk of groundwater contamination.
   The aquifer thickness of China's 37 hazardous waste landfills is clustered by the SPSS software with cluster analysis method. As shown in Table 6.9, the thickness is divided into three categories based on the cluster result. The category varies with the color.
   It is divided into three levels based on the value characteristics of aquifer depth in different categories and their positive correlations with groundwater contamination risk. See Table 6.10.
   The methods of classifying the ranking boundary of the same category of indexes are similar.
2. Depth of groundwater
   In the case of deep groundwater, it costs a long time for the poisonous and hazardous leachate from landfill to arrive at the aquifer. The adsorption and degradation process can reduce the concentration of pollutants, thereby decreasing the risk of contamination caused by groundwater. Conversely, in

Table 6.9 Cluster result for aquifer thickness of hazardous waste landfill

No.	Thickness of the aquifer (m)	No.	Thickness of the aquifer (m)	No.	Thickness of the aquifer (m)
1	8.6	14	1.31	27	2.9
2	3.3	15	7.7	28	15
3	19	16	16.6	29	4.5
4	3	17	8	30	5
5	7	18	16	31	15.2
6	5	19	3.4	32	5
7	13.5	20	24	33	9.39
8	15	21	20	34	20
9	20	22	9.39	35	20
10	21.84	23	5.2	36	27
11	6	24	2.4	37	9
12	20	25	9		
13	5.8	26	8.2		

Table 6.10 Ranking result for aquifer thickness of hazardous waste landfill

	Primary	Secondary	Tertiary
Thickness of the aquifer (m)	>18	10–18	<10

the event of shallow groundwater, the time for the leachate arriving at the aquifer is short. The extent of adsorption and degradation is relatively low. Therefore, the concentration of poisonous and hazardous substances in the leachate is relatively high and the risk of groundwater contamination is comparatively great.

The groundwater depth of China's 37 hazardous waste landfills is clustered by the SPSS software with cluster analysis method. As shown in Table 6.11, the groundwater depth is divided into three categories based on the cluster result. The category varies with the color.

It is divided into three levels based on the value characteristics of groundwater depth in different categories and their positive correlations with groundwater contamination risk. See Table 6.12.

The methods of classifying the ranking boundary of the same category of indexes are similar.

The groundwater contamination risk ranking indexes of hazardous waste landfills are provided with single-index cluster analysis, respectively, to obtain the ranking boundary of each index. See Table 6.13.

Table 6.11 Cluster result for groundwater depth of hazardous waste landfill

No.	Depth of groundwater (m)	No.	Depth of groundwater (m)	No.	Depth of groundwater (m)
1	0.5	14	1.55	27	5
2	9.7	15	4.7	28	3.8
3	126	16	20	29	1
4	2.4	17	3.2	30	50
5	7.5	18	8	31	5
6	2.5	19	7.97	32	3.8
7	5.3	20	3	33	4.1
8	19.5	21	81.5	34	1
9	20	22	3	35	1.2
10	40	23	30	36	13
11	2.5	24	50	37	1
12	25	25	2.4		
13	10	26	50		

Table 6.12 Ranking result for groundwater depth of hazardous waste landfill

	Primary	Secondary	Tertiary
Depth of groundwater (m)	<15	15–30	>30

Table 6.13 Single-index ranking result for groundwater contamination of hazardous waste landfill

	Primary	Secondary	Tertiary
Depth of groundwater (m)	<15	15–30	>30
Amount of the landfills (t/a)	>50000	20000–50000	<20000
Area of landfill (m^2)	>70000	20000–70000	<20000
Life of landfill (a)	>20	15–20	<15
Thickness of HDPE (mm)	<3	3–3.5	>3.5
Amount of leachate (m^3/d)	>30	15–30	<15
Distance (m)	<1500	1500–3000	>3000
Thickness of clay liner (cm)	<75	75–125	>125
Amount of net recharge (mm)	>250	150–250	<150
Terrain slope	>10 %	5–10 %	<5 %
Thickness of the vadose zone (m)	<4	4–8	>8
Permeability coefficient of the vadose zone (m/d)	>0.1	0.01–0.1	<0.01
Thickness of the aquifer (m)	>18	10–18	<10
Permeability coefficient of the aquifer (m/d)	>1	0.01–1	<0.01

Table 6.14 Comprehensive rating of groundwater contamination risk of China's hazardous waste landfills

No.	Grade	No.	Grade	No.	Grade
1	29.86	14	32.56	27	34.06
2	28.56	15	26.14	28	33.38
3	26.78	16	34.48	29	37.3
4	31.78	17	29.5	30	22.2
5	29.7	18	37.3	31	29.2
6	33.12	19	37.44	32	35.18
7	39.14	20	27.22	33	30.66
8	22.32	21	28.08	34	31.08
9	35.92	22	29.28	35	29.82
10	33.9	23	32.22	36	37.76
11	32.2	24	27.84	37	31.78
12	36.9	25	27.48		
13	28.9	26	24.42		

Table 6.15 Final ranking result for groundwater contamination risk of China's hazardous waste landfills

Rank	Comprehensive rating
Primary	>35
Secondary	30–35
Tertiary	<30

II. The groundwater contamination in each hazardous waste landfill is calculated for value-at-risk and then ranked.

On the basis that the ranking boundary of each index is determined with cluster analysis method and the weight of each index to overall objective is calculated through analytic hierarchy process, China's 37 hazardous waste landfills are graded comprehensively with the following methods: the index subject to primary risk is graded with 50, that subject to secondary risk is graded with 30 and that subject to tertiary risk is graded with 10. The comprehensive rating of hazardous waste landfills is calculated by weighting. See Table 6.14.

Similarly, the comprehensive rating of groundwater contamination risk of China's 37 hazardous waste landfills is clustered by the SPSS software with cluster analysis method. As shown in Table 6.15, the comprehensive rating is divided into three categories based on the cluster result.

6.3 Ranking Management of Groundwater Contamination in Landfill

6.3.1 Connotation of Risk Ranking Management

The risk ranking management (RRM) is to classify different risk ranks on the basis of the risk assessment result in accordance with confirmed procedures and methods, and also determine the scientific and reasonable risk ranking management programs and measures based on relevant technical specifications or standards in combination with social, economic, and political factors to configure the limited management resources efficiently and reasonably, give prominence to the key points, and increase the management level of risk prevention.

The risk ranking management aims to collocate the management resources reasonably. The risk management is effective for potential environment problems. The risk ranking management is an important part of the environmental risk management technology system. It must consume many unnecessary management resources to use the unified risk management method and system for the same type of management objective. The risk ranking management provides scientific basis for risk management, including (1) giving defined and quantized risk rank classification method; (2) giving the management objective that has significantly potential environmental risk to remind the decision-maker or manager to strengthen supervision as a main point; (3) giving the management objective that is in the potential environmental risk or in the low level to inform the decision-maker or manager of reducing supervision properly so as to save the management cost and increase the risk management level.

The risk ranking management started at the stage of management objective formation. The classification of risk rank may change, with the development of management objective because of surrounding environment variation, technical level as well as political, legal, social, and economic factors.

In recent years, the quantity of China's landfills is significantly increasing. They bring many difficult problems for decision-makers in the aspect of siting, construction, using, and closing. The potential risk of landfill to groundwater contamination is the particularly main problem that the decision-makers have to face during management after siting, construction, and closing. The existing specifications and standards for landfill siting and construction in China effectively prevent some contamination events. However, the standardized management method and system which is nationally uniform may result in excessive or insufficient management for special landfills. Some landfills that have significantly potential contamination risks are not supervised, especially while some others that have relatively low risks are provided with excessive management resources, which wastes the management resources. Therefore, the establishment of landfill groundwater contamination risk ranking management technology has important significance to the improvement of China's landfill risk management level and the scientific and reasonable configuration of limited management resources.

6.3.2 Procedures and Methods of Ranking Management

6.3.2.1 Principle of Ranking Management

The overall objective of landfill groundwater contamination risk ranking management is to reasonably configure the human, material, and financial resources, realize the scientific ranking and effective management of landfill groundwater contamination of China, ensure the key supervision for landfills that have significantly potential groundwater contamination accident occurrence probability, and reduce the risk of landfills to surrounding groundwater environment to the greatest extent. The following principles shall be obeyed for the ranking management:

1. Whole-process supervision and management
   The landfill groundwater contamination risk management is an enormous systems engineering. All processes shall be supervised and managed scientifically from siting, leachate collecting and disposal, landfill seepage control measures to contamination sources control, path blocking, and contamination restoration after the groundwater is polluted. No process shall be omitted.
2. Quantitative assessment and ranking management
   The risk level shall be assessed quantitatively in consideration of the potential hazards of landfill to groundwater with the risk assessment method. On this basis, the landfill groundwater contamination risk is ranked and managed.
3. Dynamic management
   The rank of risk is not consistent. If the risk of landfill to surrounding groundwater is greatly reduced after proper management, it can be classified into lower and safer rank based on the risk level for management.
4. Social management cost minimization
   The social management cost minimization is the basic principle that shall be obeyed by ranking management and also the core problem of the landfill groundwater contamination risk management program.

6.3.2.2 Ranking Management Procedure

The ranking management of landfill groundwater contamination risk shall obey the following procedure. First, the basic information such as the hydrogeological conditions of landfill and surroundings shall be mastered on the basis of data collection and site survey. Then, the landfill groundwater contamination risk is ranked with the ranking method established in this chapter. Finally, the targeted risk ranking management program is established on the above basis. The risk ranking management program for landfills that have been completed and put into operation shall be established in the aspects of landfill control, landfill seepage control system restoration, and groundwater contamination restoration. Additionally, the risk ranking management program for landfills the have not been completed or put into operation shall be established in the aspects of siting, standardized construction, and so on. Then, the groundwater is monitored. If the status is improved, it suggests that the established ranking management program is effective and feasible. Otherwise, the program needs to be regulated.

6.3.2.3 Ranking Management Program

The following ranking management program is established based on the above landfill groundwater contamination risk ranking management procedures to realize the unification of management program effectiveness, pollution treatment project environment safety and economic feasibility, prevent the migration and spread of pollutants, and guarantee the groundwater using function.

1. Landfills that have been completed and put into operation

Primary risk:

(a) The landfill operation shall be stopped immediately. The landfill seepage control system shall be provided with leakage detection. The impermeable layer on the part that is leaking shall be restored. The contamination sources shall be controlled.
(b) The mobilized wastes shall be controlled in strict accordance with relevant specifications of *Standard for Pollution Control on the Landfill Site of Municipal Solid Waste* (GB16889–2008) and *Standard for Pollution Control on the Security Landfill Site for Hazardous Wastes* (GB18598–2001). The landfill operation management requirements and pollution control requirements shall be strictly obeyed.
(c) The polluted groundwater shall be restored with effective restoration method based on the landfill groundwater contamination extent and particular pollutants to resume its using function.
(d) More efforts shall be paid to the groundwater monitoring of landfill. The groundwater shall be sampled once a month for monitoring. Key attention shall be paid to three nitrogen, heavy metal, and organic pollutants until all the monitoring indexes of groundwater are below the three-level criteria of *Quality Standard for Groundwater* (GB14848–93).

Secondary risk:

(a) The landfill seepage control system shall be provided with leakage detection. The impermeable layer on the part that is leaking shall be restored. The contamination sources shall be controlled.
(b) The mobilized wastes shall be controlled in strict accordance with relevant specifications of *Standard for Pollution Control on the Landfill Site of Municipal Solid Waste* (GB16889–2008) and *Standard for Pollution Control on the Security Landfill Site for Hazardous Wastes* (GB18598–2001). The landfill operation management requirements and pollution control requirements shall be strictly obeyed.
(c) More efforts shall be paid to groundwater monitoring of landfill properly. The groundwater shall be sampled every two months for monitoring. The key attention shall be paid to three nitrogen, heavy metal, and organic pollutants. The effective restoration means shall be taken to restore pollution in the case of excessive pollutants.

Tertiary risk:

(a) The mobilized wastes shall be controlled in strict accordance with relevant specifications of *Standard for Pollution Control on the Landfill Site of Municipal Solid Waste* (GB16889–2008) and *Standard for Pollution Control on the Security Landfill Site for Hazardous Wastes* (GB18598–2001). The landfill operation management requirements and pollution control requirements shall be strictly obeyed.
(b) The groundwater shall be sampled for monitoring quarterly according to normal groundwater monitoring frequency. The key attention shall be paid to three nitrogen, heavy metal, and organic pollutants. The effective means shall be taken to control the contamination sources or restore pollution in the case of excessive pollutants.

2. Solid waste disposal sites under planning and construction

Primary risk:

The landfill shall be sited again in accordance with landfill siting requirements of *Standard for Pollution Control on the Landfill Site of Municipal Solid Waste* (GB16889–2008) and *Standard for pollution control on the security landfill site for hazardous wastes* (GB18598–2001). The landfill shall be constructed according to the landfill design and construction requirements. Special attention shall be paid to the impermeable layer as well as the design and construction of leachate guide and drainage system. The emission electrode and receiving electrode shall be installed, respectively, in the landfill and in the soil near surface. The electrode grid shall be installed below the geomembrane during construction so as to detect the leakage position, size, and quantity of impermeable layer with electrical method in the process of landfill operation.

Secondary risk:

The landfill shall be constructed in accordance with the landfill design and construction requirements of *Standard for Pollution Control on the Landfill Site of Municipal Solid Waste* (GB16889–2008) and *Standard for pollution control on the security landfill site for hazardous wastes* (GB18598–2001). The design and construction of the impermeable layer and the leachate guide and drainage system shall be strengthened properly. The emission electrode and receiving electrode shall be installed, respectively, in the landfill and the soil near surface. The electrode grid shall be installed below the geomembrane during construction so as to detect the leakage position, size, and quantity of impermeable layer with electrical method in the process of landfill operation.

Tertiary risk:

The landfill shall be constructed based on the preset program. All work shall be conducted in strict accordance with relevant provisions of *Standard for Pollution Control on the Landfill Site of Municipal Solid Waste* (GB16889–2008) and *Standard for pollution control on the security landfill site for hazardous wastes*

(GB18598–2001). The emission electrode and receiving electrode shall be installed, respectively, in the landfill and the soil near surface. The electrode grid shall be installed below the geomembrane during construction so as to detect the leakage position, size, and quantity of impermeable layer with electrical method in the process of landfill operation.

References

Chen NX, Dong GM, He XC. Fuzzy comprehensive evaluation of groundwater environmental vulnerability based on AHP. J North China Inst Water Conservancy Hydroelectric Power. 2005;26:63–6.

Canadian Council of Ministers of the Environment. National Classification System for Contaminated Sites. Report CCME EPC-CS39E Winnipeg, Canada;2008.

Index

B. Xi et al., *Optimization of Solid Waste Conversion Process and Risk Control of Groundwater Pollution*, SpringerBriefs in Environmental Science,
DOI 10.1007/978-3-662-49462-2

O

P

Q

R

S

T

V

W

Printed in the United States
By Bookmasters